NIST Measurement Services:
Gas Flowmeter Calibrations with the 26 m³ *PVTt* Standard

NIST Special Publication 250-1046

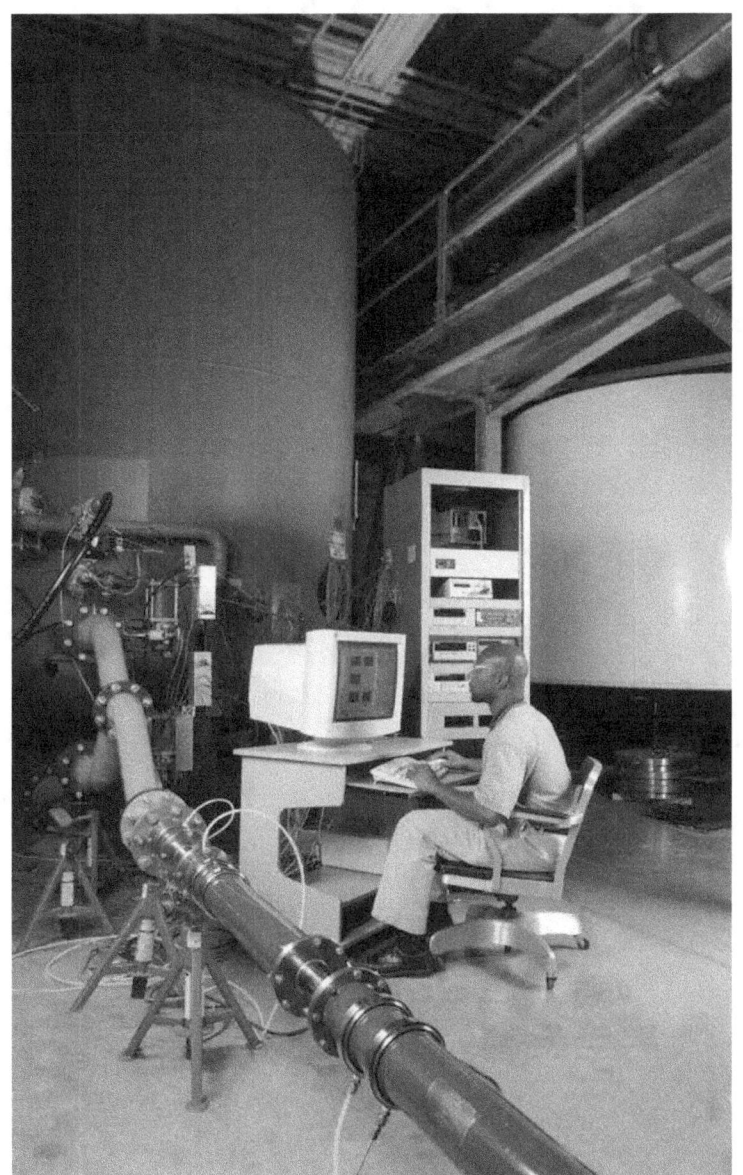

Aaron N. Johnson and John D. Wright
November 25, 2009

U. S. Department of Commerce
Technology Administration
National Institute of Standards and Technology

Table of Contents Gas Flowmeter Calibrations with the 26 m³ *PVTt* Standard

Abstract 1
1 Introduction to Gas Flow Measurement at NIST 2
2 Description of Gas Flow Calibration Services 3
3 Procedures for Submitting a Flowmeter for Calibration 5
4 Overview of Pressure, Volume, Temperature, and time (*PVTt*) Flow Standards 5
    4.1 Description of the 26 m³ *PVTt* Flow Standard 5
    4.2 CFV Check Standards 6
    4.3 Theoretical Development of the *PVTt* Mass Flow 7
    4.4 *PVTt* Operating Procedures 9
    4.5 Inventory Volume Mass Cancellation Technique 11
5 Uncertainty of *PVTt* Subsidiary Components 14
    5.1 Reference Parameters (Molec. Weight, Univ. Gas Const, Comp. Factor) 15
    5.2 Collection Time 16
    5.3 Pressure and Temperature in the Inventory Volume 17
    5.4 Inventory Volume 22
    5.5 Pressure and Temperature in the Collection Tank 22
    5.6 Collection Tank Volume 25
6 Mass Flow Uncertainty 32
    6.1 Accumulated Mass in Collection Tank 33
    6.2 Accumulated Mass in Inventory Volume 34
    6.3 Effect of Leaks 36
    6.4 Uncertainty Attributed to the Steady Flow Assumption 37
7 Summary 37
8 References 38
    Appendix: Sample Calibration Report

Abstract

This document describes NIST's 26 m^3 pressure, volume, temperature, and time (*PVTt*) primary flow standard. This standard is used to calibrate gas flow meters over a range extending from 200 L/min to 77000 L/min where the reference temperature and pressure conditions are 293.15 K and 101.325 kPa respectively, and the working fluid is dry air. This standard measures flow by collecting a steady stream of gas into a tank of known volume during a measured time interval. The ratio of the mass of gas accumulated in the tank to the collection time is the mass flow.

The *PVTt* standard measures mass flow with an expanded uncertainty of 0.09 % at the 95 % confidence interval (*i.e.*, $k = 2$) over the full flow range. This document presents a detailed uncertainty analysis evaluating and explaining the various components that comprise the expanded uncertainty. In addition to the uncertainty analysis, we specify the various components of the 26 m^3 *PVTt* system, describe its theoretical basis of flow measurement, document the calibration procedure used for the standard, provide information about the calibration service (*i.e.*, meters that we commonly test, available pipe sizes, sample calibration report, etc.), and give details regarding the unique aspects of NIST's *PVTt* systems.

Key words: calibration, uncertainty, flow, flowmeter, gas flow standard, inventory volume, *PVTt* standard, inventory mass cancellation technique, sensor response, correlated uncertainty sources.

1. Introduction to Gas Flow Measurement at NIST

Calibrations of gas flow meters are performed with primary standards [1] that are based on measurements of more fundamental quantities, such as length, mass, and time. Primary flow calibrations are accomplished by collecting a measured mass or volume of a flowing fluid over a measured time interval. The ratio of the collected mass to the measured time interval equals the time-averaged mass flow at the meter under test (MUT). To ensure that the instantaneous mass flow equals the time-averaged value, the flow at the MUT should be maintained under steady state conditions of flow, pressure, and temperature.

Traditionally, primary flow standards have been based on either gravimetric or volumetric methods. Gravimetric based primary flow standards measure the mass of collected gas by directly weighing the mass of the collection vessel before and after gas accumulation [2]. On the other hand, volumetric based primary standards calculate the mass of collected gas by multiplying the measured density of the gas by the volume of the collection tank. Common volumetric based primary standards include piston provers [3], bell provers [4], and pressure-volume-temperature-time (*PVTt*) systems [5, 6]. In the Fluid Metrology Group (FMG) at NIST, gas flow is measured exclusively with *PVTt* primary flow standards.

Table 1. Flow measurement capabilities of the three NIST *PVTt* primary gas flow standards.

Flow Standard	Flow Range (L/min)	Gas Type	Pressure Range (kPa)	Relative Expanded Uncertainty ($k = 2$) (%)
34 L *PVTt*	1 to 100	Dry Air	100 to 1700	0.05
	1 to 100	N_2	100 to 7000	0.03 to 0.04
	1 to 100	CO_2	100 to 4000	0.05
	1 to 100	Ar	100 to 7000	0.05
	1 to 100	He	100 to 7000	0.05
677 L *PVTt*	10 to 2000	Air	100 to 1700	0.05
	10 to 150	N_2	100 to 800	0.02 to 0.03
26 m^3 *PVTt*	200 to 77000	Dry Air	200 to 800	0.09

The FMG of the Process Measurements Division (part of the Chemical Science and Technology Laboratory) at NIST has three *PVTt* flow standards that provide gas flow calibration services over a range from 1 L/min to 77000 L/min.[1] The lowest flows are measured using the 34 L *PVTt* system, medium flows using the 677 L *PVTt* system, and the largest flows using the 26 m^3 *PVTt* system. The various flow ranges, types of gases, pressure ranges, and uncertainty for each *PVTt* primary standard are shown in Table 1. (Not included in this table are flows less than 1 L/min, which can be calibrated by the NIST Pressure and Vacuum Group.)

This document discusses the procedures for submitting a flow meter for calibration, gives the readily available pipe sizes and flanges suitable for flow meter calibration, documents the format of a standard NIST calibration report, and states the normal range of data collected for a calibration. In addition, this document describes the theory, principle of operation, and

---

[1] Reference conditions of 293.15 K and 101.325 kPa are used throughout this document for volumetric flows.

uncertainty of the 26 m^3 *PVTt* primary flow standard covering the flow range from 200 L/min to 77000 L/min. Details concerning the two smaller *PVTt* flow standards can be found in the following reference [6].

2. Description of Gas Flow Calibration Services

NIST offers calibrations of gas flow meters in order to provide traceability to flow meter manufacturers, secondary flow calibration laboratories, and flow meter users. For a calibration fee, NIST calibrates a customer's flow meter and delivers a calibration report that documents the calibration procedure, the calibration results, and their uncertainty. The flow meter and its calibration results may be used in different ways by the customer. The flow meter is often used as a transfer standard to perform a comparison of the customer's primary standards to the NIST primary standards so that the customer can establish traceability, validate their uncertainty analysis, and demonstrate proficiency. Customers with no primary standards frequently use their NIST calibrated flow meters as working standards or reference standards in their laboratory to calibrate other flow meters.

Table 2. Readily available pipe sizes and fittings. [2]

Nominal Pipe Diameter		Fittings and/or ANSI Flange Ratings
(cm)	(in)	
2.54	1	VCO, Swagelok, AN, and NPT
5.08	2	ANSI Flanges 150 and 300
7.62	3	ANSI Flange 300
10.16	4	ANSI Flanges 150 and 300
15.24	6	ANSI Flange 300
20.32	8	ANSI Flanges 300 and 600

Flowmeters can be calibrated in pipe sizes ranging from 2.54 cm (1 in) to 20.32 cm (8 in). The standard pipe sizes and flanges used in the 26 m^3 *PVTt* flow standard are listed in Table 2. Flow meters can be tested if the flow range, gas type, and piping connections are suitable, and if the system to be tested has precision appropriate for calibration with the NIST flow measurement uncertainty. The vast majority of flow meters calibrated in the gas flow calibration service are critical flow venturis (CFVs), also commonly called critical nozzles. To date, this meter type is regarded as the best candidate for transfer and working standards by the gas flow metrology community [7]. Other meter types that we have tested include laminar flow meters, positive displacement meters, roots meters, rotary gas meters, thermal mass flow meters, and turbine meters. Meter types with precisions or calibration instabilities that are significantly larger than the uncertainty of the primary standard should not be calibrated by NIST for economic reasons. For example, a rotameter for which the float position is read by the operator's eye normally

---

[2] Certain commercial equipment, instruments, or materials are identified in this paper to foster understanding. Such identification does not imply recommendation or endorsement by the National Institute of Standards and Technology, nor does it imply that the materials or equipment identified are necessarily the best available for the purpose.

cannot be read with precision any better than 1 %. It is not practical to pay several thousand dollars to obtain a NIST calibration with an expanded uncertainty of 0.09 %. For such a flowmeter, a calibration with an expanded uncertainty of 0.5 % would be perfectly adequate and is available from other laboratories at significantly lower cost.

A normal flow calibration performed by the NIST Fluid Flow Group consists of five flows spread over the range of the flow meter. For a CFV, typical calibration set points are at 200 kPa, 300 kPa, 400 kPa, 500 kPa, and 600 kPa. A laminar flow meter is normally calibrated at 10 %, 25 %, 50 %, 75 %, and 100 % of the meter full scale. At each of these flow set points, three (or more) flow measurements are made with the *PVTt* standard. The same set point flows are tested on a second occasion, but the flows are tested in decreasing order instead of the increasing order of the first set. Therefore, the final data set consists of six (or more) primary flow measurements made at five flow set points (*i.e.,* at least 30 individual flow measurements). The sets of three measurements can be used to assess repeatability, while the sets of six can be used to assess reproducibility. For further explanation, see the sample calibration report that is included in this document as an appendix. Variations on the number of flow set points, spacing of the set points, and the number of repeated measurements can be discussed with the NIST technical contacts. However, for data quality assurance reasons, we rarely will conduct calibrations involving fewer than three flow set points and two sets of three flow measurements at each set point.

The FMG prefers to present flow meter calibration results in a dimensionless format that takes into account the physical model for the flow meter type. The dimensionless approach helps facilitate accurate flow measurements by the flow meter user even when the conditions of usage (*i.e.,* gas type, temperature, pressure) differ from the conditions during calibration. For example, for a CFV calibration, the calibration report will present Reynolds number and discharge coefficient, and for a laminar flow meter, a report presents the viscosity coefficient and the flow coefficient [8]. However, we point out that there may be additional uncertainties introduced when the dimensionless approach is used to *extrapolate* a NIST calibration to conditions or gases that were not specifically tested. Unless special provisions are made between NIST and the customer, it is the customer's responsibility to determine any additional uncertainty when using the dimensionless approach to extend a NIST calibration beyond the measured range.

When a flow meter is calibrated, the uncertainty of its dimensionless calibration factors depend on both the uncertainty of the flow standard as well as the uncertainty of the instrumentation associated with the MUT (normally absolute pressure, differential pressure, and temperature instrumentation). We prefer to connect our own instrumentation (temperature, pressure, etc.) to the meter under test since they have established uncertainty values based on calibration records that we would not have for the customer's instrumentation. In some cases, it is impractical to install our own instrumentation on the MUT. This situation typically occurs when the MUT outputs flow. In this case, we provide a table of flow indicated by the MUT, flow measured by the NIST standard, and the uncertainty of the NIST flow value.

Customers should consult the web address http://ts.nist.gov/MeasurementServices/Calibrations/mechanical_index.cfm to find the most current information regarding our calibration services, calibration fees, technical contacts, and flow meter submittal procedures.

3. Procedures for Submitting a Flow meter for Calibration

The FMG follows the policies and procedures described in Chapters 1, 2, and 3 of the NIST Calibration Services Users Guide [9]. These chapters can be found on the internet at the following address:
http://ts.nist.gov/MeasurementServices/Calibrations/mechanical_index.cfm
Chapter 2 gives instructions for ordering a calibration for domestic customers and has the sub-headings: A.) Customer Inquiries, B.) Pre-arrangements and Scheduling, C.) Purchase Orders, D.) Shipping, Insurance, and Risk of Loss, E.) Turnaround Time, and F.) Customer Checklist. Chapter 3 gives special instructions for foreign customers.

4. Overview of Pressure, Volume, Temperature, and time (*PVTt*) Flow Standards

NIST has used *PVTt* systems as a primary gas flow standards for more than 30 years [5, 6]. In this section we provide an overview of the NIST *PVTt* facility, develop its theoretical basis for flow measurements, list its operating procedures, and detail its unique features that distinguish it from *PVTt* systems used in other laboratories.

4.1 Description of NIST 26 m^3 *PVTt* System

The NIST 26 m^3 *PVTt* calibration system is the United States primary standard for measuring gas flows ranging from 200 L/min to 77000 L/min. The relative expanded uncertainty over this range of flows is 0.09 % ($k = 2$). The working fluid is filtered, dry air supplied by a three stage centrifugal compressor in series with a desiccant drier. The compressor delivers airflow at line pressures up to 800 kPa at nominally room temperature conditions and at relative humidity levels below 3 %.

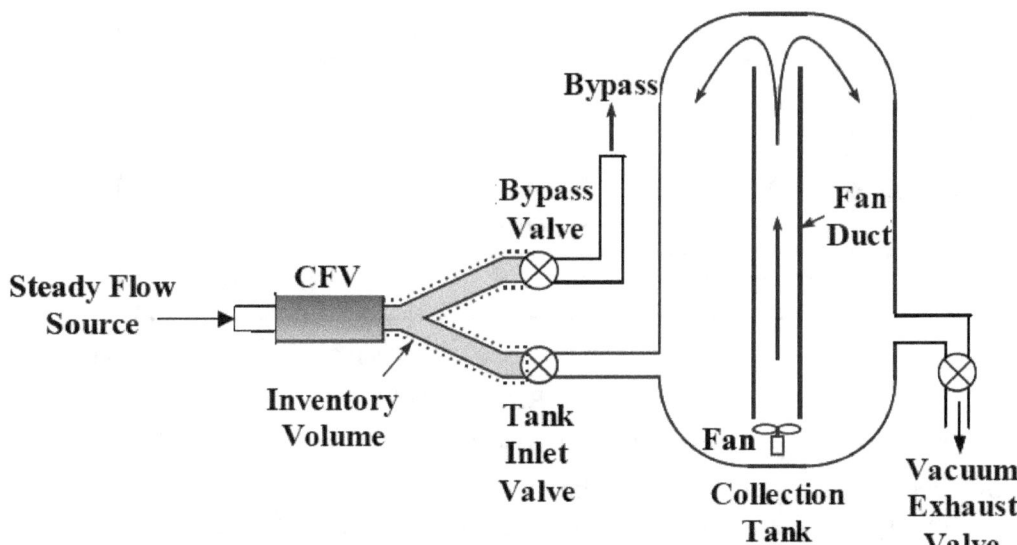

Figure 1. Schematic diagram of the NIST 26 m^3 *PVTt* gas flow standard.

Flow measurements using the 26 m^3 *PVTt* flow standard are completely automated using LabVIEW[2] software. This software controls each facet of the calibration process including setting the nominal flow, actuating the valves, filling and evacuating the collection tank, taking the appropriate pressure and temperature data, measuring the collection time interval, and reducing the data. The post-processed calibration data that is calculated by the LabVIEW program is verified by recalculating the data on a spreadsheet. The *PVTt* system is equip with various safety features that prevent overpressurizing the collection tank during a calibration. This allows the *PVTt* system to safely perform calibrations during non-business hours, thereby allowing a faster turnaround time for our customers.

The main components of the *PVTt* calibration system include a source of steady flow, a set of appropriately sized CFVs to cover the flow range, an inventory volume sized appropriately for the flow, the collection tank, a timing mechanism, a data acquisition system, and pressure and temperature instrumentation. A schematic of the *PVTt* system showing some of these components is depicted in Fig. 1. The inventory volume functions to divert the flow to either the collection tank or bypass. The timing system measures the duration that gas accumulates in the collection tank and inventory volume. The collection tank stores the gas, allowing it to thermally equilibrate before determining its mass. The CFV plays multiple roles. First, it isolates the steady upstream flow at the CFV inlet from downstream pressure fluctuations that occur in the inventory volume during actuation of the bypass and tank inlet valves. Second, the sonic line at the CFV throat, in conjunction with the bypass and tank inlet valves, provides a definite boundary for the inventory volume. Lastly, it serves as a check standard to help ensure that the *PVTt* system performs consistently over time.

4.2 CFV Check Standards

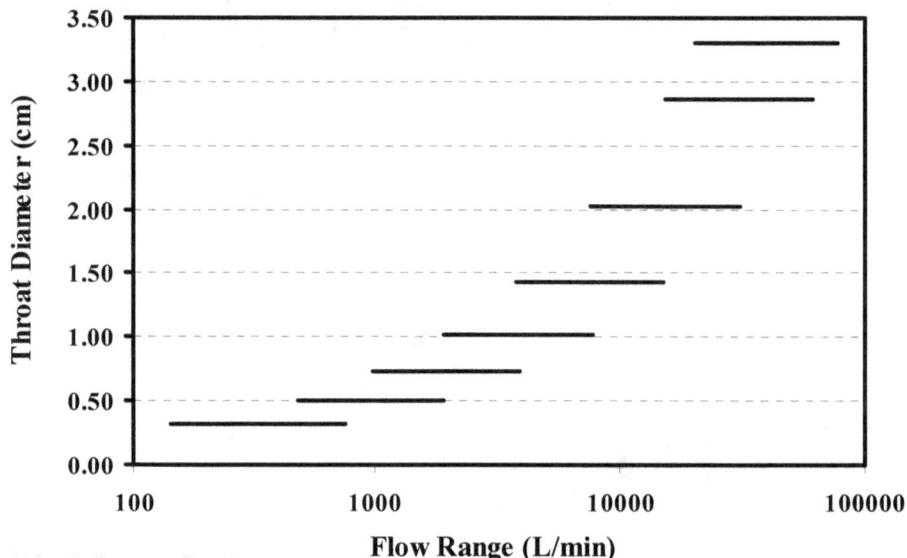

Figure 2. Flow ranges covered by various sized CFVs calibrated on the NIST 26 m^3 *PVTt* gas flow standard.

Critical flow venturis are considered to be among the best transfer standards by the flow metering community and are commonly used as transfer standards for international comparisons between National Metrology Institutes [7]. Since these devices are inherently part of a *PVTt* system, the FMG maintains a calibrated set of variously sized CFVs that span the flow range of its three *PVTt* flow standards (*i.e.*, 34 L, 677 L, and 26 m^3 *PVTt* standards). These CFVs are used as working standards in another calibration facility called the Working Gas Flow Standard (WGFS). The WGFS provides calibrations, particularly for laminar flow meters, in which the reference flow is measured with a relative expanded uncertainty no greater than 0.16 % ($k = 2$).

Figure 2 shows the subset of WGFS CFVs that are calibrated on the 26 m^3 *PVTt* flow standard. The sizes of these CFVs were selected so that the lower and upper limits of their flow ranges overlap. The smallest three CFVs in Fig. 2 have a portion of their flow ranges that can be calibrated on both the 677 L and the 26 m^3 *PVTt*. Agreement between these independent systems adds confidence to the validity of each system's calibration results.

4.3 Theoretical Development of the *PVTt* Mass Flow

PVTt systems measure the CFV mass flow[3] using timed-collection techniques based on the principle of conservation of mass. For fluid flow into an arbitrary control volume (*i.e.*, region of interest), this principle requires that the rate of mass accumulation in the control volume equals the net influx of mass through its boundaries. In Fig. 1 we take the control volume to include both the collection tank and the inventory volume so that the statement of mass conservation is

$$\dot{m}_{net} = \frac{dM}{dt} \tag{1}$$

where the total mass in the control volume includes both M_T, the mass in the collection tank, and M_I, the mass in the inventory volume,

$$M = M_T + M_I \tag{2}$$

and the net influx of mass into the control volume is

$$\dot{m}_{net} = \dot{m} + \dot{m}_{leak} \tag{3}$$

the summation of the CFV mass flow, $\dot{m}$, and the leakage of mass flow into the control volume from the environment surrounding the tank, $\dot{m}_{leak}$. Although *PVTt* systems are designed to measure $\dot{m}$, they do not distinguish between the CFV mass flow and flow from other sources (*i.e.*, leaks), and therefore the flow that is actually measured is $\dot{m}_{net}$. Consequently, $\dot{m}_{leak}$ must be either known or negligibly small relative to $\dot{m}$.

The effects of leaks can be understood by Eqn. (3), which shows that leakage into the control volume will result in overpredicting the actual mass flow (*i.e.*, $\dot{m}_{net} > \dot{m}$). Conversely, the actual mass flow will be underpredicted (*i.e.*, $\dot{m}_{net} < \dot{m}$) for leakage out of the control volume. During a calibration cycle, the gas pressures inside the collection tank and inventory volume[4] are maintained at or below atmospheric pressures so that leaks tend to flow into the control volume, causing the flow standard to overpredict the actual mass flow. The FMG regularly inspects its flow standard for leaks to ensure the quality of calibration data. In cases where leaks cannot be completely eliminated their effects are included as part of the uncertainty analysis (see section 6.3).

The expression for mass flow as given by Eqn. (1) is not useful in its present form since the rate of mass accumulation in the control volume (*i.e.*, the derivative term), in general cannot be directly measured at low levels of uncertainty. This difficulty is circumvented by maintaining steady state conditions of pressure and temperature in the piping section upstream of the CFV

---

[3] *PVTt* systems can also be used to measure the mass flow of a MUT located upstream of the CFV. In this case the uncertainty analysis presented in this document should be modified to include the mass storage effects that occur in the piping volume between the MUT and CFV.

[4] The pressure in the inventory volume briefly exceeds one atmosphere during flow diversion, but are sub-atmospheric during the majority of the collection interval.

inlet. As long as the appropriate pressure ratio is maintained across the CFV, the mass flow ($\dot{m}$) remains constant throughout the collection period. If the leak rate is negligible, then the *instantaneous* rate of mass accumulation, $\frac{dM}{dt}$, is constant, and equals

$$\frac{dM}{dt} = \frac{\Delta M}{\Delta t} \qquad (4)$$

the *average* rate where $\Delta t = t^f - t^i$ is the collection period, and $\Delta M = M^f - M^i$ is the mass accumulated during this period. Here, the initial and final masses in the control volume, M^i and M^f, correspond to the times coinciding with the start and end of the collection period, t^i and t^f, respectively. The total accumulated mass in the control volume consist of ΔM_T, mass accumulated in the collection tank, and ΔM_I, the mass accumulated in the inventory volume

$$\Delta M = \left(M_T^f - M_T^i\right) + \left(M_I^f - M_I^i\right) = \Delta M_T + \Delta M_I \qquad (6a)$$

Each of the four of the masses in Eqn. (5) are determined by multiplying the appropriate volume (either the collection tank or inventory volume) by the average gas density at the time of interest. Both the collection tank and inventory volumes are determined prior to a calibration cycle. They are measured as described in sections 5.4 and 5.6 respectively. If both volumes are assumed to remain fixed over the range of temperatures and pressures they experience, the mass accumulation in the collection tank and inventory volumes are[5]

$$\Delta M_T = \left(\rho_T^f - \rho_T^i\right) V_T \qquad (6a)$$

$$\Delta M_I = \left(\rho_I^f - \rho_I^i\right) V_I \qquad (6b)$$

where V_T and V_I are the respective collection tank and inventory volumes. Applying the equation of state for gas density, $\rho = P\mathcal{M}/ZR_u T$, the accumulated masses in the collection tank and in the inventory volume are

$$\Delta M_T = (\mathcal{M}/R_u)\left(\frac{P_T^f}{Z_T^f T_T^f} - \frac{P_T^i}{Z_T^i T_T^i}\right) V_T \qquad (7a)$$

$$\Delta M_I = (\mathcal{M}/R_u)\left(\frac{P_I^f}{Z_I^f T_I^f} - \frac{P_I^i}{Z_I^i T_I^i}\right) V_I \qquad (7b)$$

where $\mathcal{M}$ is the molecular weight of the dry air [10], R_u is the universal gas constant [11], Z is the compressibility factor for dry air [10], and P and T are the average pressure and temperature, respectively. By combining Eqns. (4) and (5) and substituting the result into Eqn. (1) the governing expression for mass flow is

$$\dot{m} = \frac{\Delta M_T + \Delta M_I}{\Delta t} \qquad (8)$$

---

[5] The change in the collection tank volume due its elasticity and thermal expansion between its evacuated and filled conditions makes a negligible contribution to the uncertainty in mass flow and is neglected.

where the effect of leaks is omitted in calculating the CFV mass flow, but accounted for in the mass flow uncertainty in section 6.3. Furthermore, by substituting the definitions of ΔM_T and ΔM_I given in Eqns. (7a) and (7b) into Eqn. (8), the CFV mass flow is also given by

$$\dot{m} = \left(\frac{M/R_u}{\Delta t}\right)\left[\left(\frac{P_T^f}{Z_T^f T_T^f} - \frac{P_T^i}{Z_T^i T_T^i}\right)V_T + \left(\frac{P_I^f}{Z_I^f T_I^f} - \frac{P_I^i}{Z_I^i T_I^i}\right)V_I\right]. \qquad (9)$$

4.4 *PVTt* Operating Procedures
The typical process for measuring mass flow with the 26 m³ *PVTt* flow standard entails the following procedure:
1. With the tank valve closed, open the bypass valve and establish a stable flow through the CFV at the desired stagnation pressure (see Fig. 1).
2. Evacuate the collection tank volume (V_T) to a prescribed lower pressure using the vacuum pump. (Steps 1 and 2 can begin simultaneously.)
3. Wait for pressure and temperature conditions in the tank to stabilize and then acquire their initial values (P_T^i and T_T^i). These values will be used to calculate the initial gas density in the tank (ρ_T^i) and subsequently the initial mass of gas in the tank (M_T^i). With the tank under vacuum conditions, reasonable pressure and temperature stability is attained in 300 s or less.
4. With the tank valve still closed, close the bypass valve. After the bypass is fully closed, the flow exhausting from the CFV will dead-end in the inventory volume for a brief interval (*i.e.*, 100 ms or less) called the *first dead-end interval*. The time history of the pressure and temperature in the inventory volume is measured during the dead-end interval. The start of the collection time, (t^i), is selected within this interval. The initial pressure and temperature in the inventory volume (P_I^i and T_I^i) correspond to the selected start time. These values of pressure and temperature are used with an equation of state to determine the initial compressibility factor (Z_I^i), and subsequently the initial density (ρ_I^i), which when multiplied by the inventory volume (V_I) equals the initial mass in the inventory volume (M_I^i). Immediately following the dead-end interval, the tank valve is opened.
5. Wait for the tank to fill to a prescribed upper pressure (*i.e.*, near atmospheric pressure) and close the tank valve.
6. When the tank valve is fully closed (with the bypass valve still closed) there is a brief time interval where the flow emanating from the CFV is again dead-ended in the inventory volume, the *second dead-end interval*. The time history of both the pressure and temperature in the inventory volume are again measured during this period. The pressure and temperature data are used with the equation of state to calculate the time history of the gas density. A stop time, (t^f), is selected within the second dead-end time so that the final inventory gas density equals its initial density (*i.e.*, $\rho_I^f = \rho_I^i$), and hence the final mass in the inventory volume (M_I^f) is the same as the initial mass (M_I^i). Immediately following the second dead-end interval, open the bypass valve.
7. Turn the fan on inside the collection tank and wait for temperature stability before acquiring the final pressure and temperature (P_T^f and T_T^f). These values will be used with an equation

of state for dry air to determine the final compressibility factor (Z_T^f), and subsequently the final density (ρ_T^f). The volume of the collection tank (V_T) is multiplied by the final density to determine the final mass of gas in the collection tank (M_T^f). The usual waiting period for pressure and temperature stability is 2700 s. (Steps 6 and 7 can begin simultaneously.)
8. Equation (9) is used to determine the CFV mass flow ($\dot{m}$).
9. Return to step 1 for next calibration point or end calibration.

Table 3. Nominal values of the parameters and measured variables used in Eqn. (9).

System Components and Parameters	Quantity	Nominal Value	Instrumentation or Reference
Reference Parameters	Universal Gas Constant, R_u	8134.472 J/(kg·K)	Reference [11]
	Molecular. Mass (dry-air), $\mathcal{M}$	28.9647 g/mol	Reference [10]
	Compressibility Factor (dry-air), Z	$Z = Z(P, T)$	Reference [10]
Collection Tank	Initial Pressure, P_T^i	0.08 kPa to 0.1 kPa	Vacuum Gauge
	Final Pressure, P_T^f	93 kPa to 103 kPa	Abs. Pressure Gauge
	Initial Temperature, T_T^i	292 K to 297 K	37 Thermistors
	Final Temperature, T_T^f	292 K to 297 K	
	Volume, V_T	25.8969 m^3	see section 5.6
Inventory Volume	Initial Pressure, P_I^i	100 kPa to 450 kPa	2 Fast Pressure Transducers
	Final Pressure, P_I^f	100 kPa to 450 kPa	
	Initial Temperature, T_I^i	293 K to 320 K	2 Thermocouples
	Final Temperature, T_I^f	293 K to 320 K	
	Volume, V_I	0.025 m^3 to 0.1 m^3	see section 5.4
Timing System (see Eqn. 11)	Base time, $\Delta\tau$	20 s to 8300 s	2 Universal Counters
	1st Dead-End Interval, Δt_1	0.03 to 0.1 sec	Data acquisition card sampling at 3000 Hz
	2nd Dead-End Interval, Δt_2	0.03 to 0.1 sec	

Table 3 list the instrumentation used to make the pressure, temperature and time measurements as well as their normal range of values during a calibration. The table also gives the values of the reference parameters R_u, $\mathcal{M}$, and Z. The measurement of the collection tank volume and the inventory volume are discussed later in sections 5.4 and 5.6 respectively.

4.5 Inventory Volume Mass Cancellation Technique

Many of the operating procedures used by the FMG are standard to all blow-down *PVTt* systems. However, the *inventory mass cancellation technique* outlined in steps 4 and 6 of the *PVTt* operating procedures (see section 4.4) is unique to NIST. During the dead-end periods, both the pressures and temperatures in the inventory volume increase. The start and stop times, t^i and t^f, are selected so that the initial and final densities in the inventory volume are equal. Since the size of inventory volume remains fixed for both dead-end intervals, matching the densities ensures that the accumulated mass in the inventory volume is identically zero (*i.e.*, $\Delta M_I = 0$).

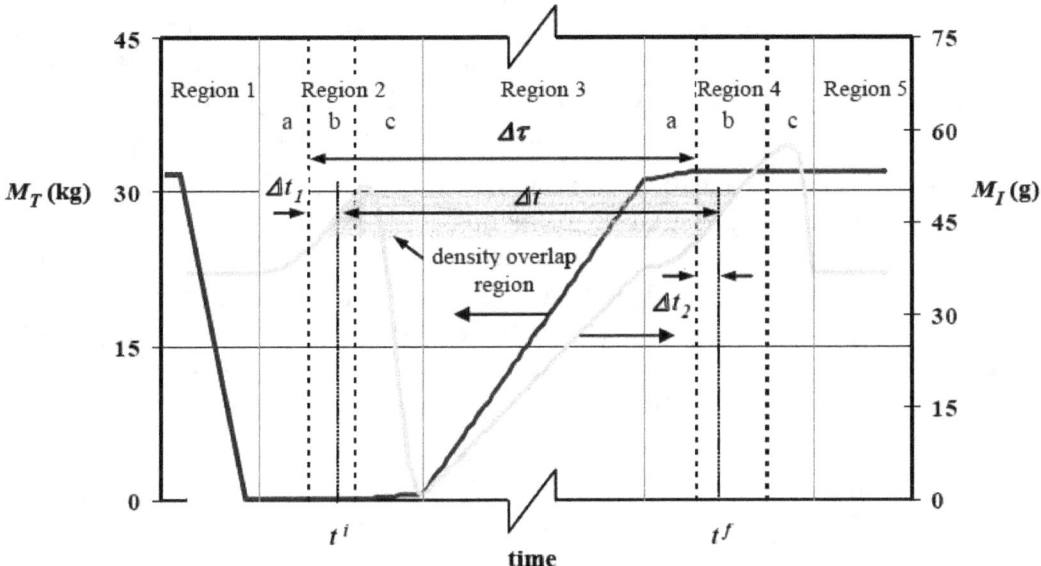

Figure 3. The time history of the mass in the inventory volume (M_I), and the mass in the collection tank (M_T) for a typical calibration cycle. (The mass plots are based on a semi-empirical model.)

The inventory volume mass cancellation technique is an extension of the pressure-matching scheme described in [6]. In the pressure-matching scheme, the initial and final pressures in the inventory volume are matched so that the accumulated mass in the inventory volume is nearly zero. The mass cancellation technique, introduced here, further develops this strategy, by matching the initial and final densities. By matching density instead of the pressure, the initial and final masses are made to completely cancel. The advantages of these matching schemes are two fold. First, the correlated uncertainty sources between the initial and final densities will completely cancel. Second, because the uncertainty in the size of inventory volume does not significantly contribute to the mass flow uncertainty, a highly accurate measurement technique is not necessary to determine the size of the inventory volume. In practice, the size of the inventory volume is rudimentarily measured to within 25 % of its actual size using a simple tape measure. This straightforward approach for measuring the size of the inventory volume is especially convenient when calibrating customer CFVs requiring modifications to the normal piping configuration of the inventory volume.

Figure 3 illustrates how the inventory mass cancellation technique is applied during a calibration cycle. The figure shows time histories for the mass in collection tank, M_T (left), and the mass in the inventory volume, M_I (right), during a typical calibration cycle. The values of M_I and M_T are obtained from a semi-empirical model based on mass conservation. The results of the model agree reasonably well with measured results, and are used here to explain the inventory matching technique.

The time histories of M_I and M_T are divided into five regions. Region 1 corresponds to steps 1 and 2 in the *PVTt* operating procedures. In this region, M_I is constant since the mass flow entering the inventory volume through the CFV equals the mass flow exiting via the bypass valve (see Fig. 1). Simultaneously, M_T decreases as the collection tank is evacuated via the vacuum exhaust valve. Region 2 corresponds to step 4 where the flow is diverted from the bypass into the collection tank. Region 3 includes the first part of step 5 where flow accumulates in the collection tank through the tank inlet valve (bypass is closed). The latter part of step 5, and step 6 correspond with Region 4 where the flow is diverted from the tank back to the bypass. Finally, Region 5 corresponds to the end of the calibration cycle as explained in step 9.

The time durations of Regions 2 and 4, corresponding to flow diversion into and away from the collection tank, have been expanded relative to the other regions in Fig 3. These brief intervals play an important role in the mass cancellation technique. By expanding these regions, the behavior of M_I can be clearly identified. Region 2 and 4 each last approximately 0.3 s, in contrast to Region 3, which can last from 20 s to 5500 s depending on flow, and Regions 1 and 5 which together, last approximately 4000 s. Regions 2 and 4 are both divided into three distinct subdivisions labeled "a", "b", and "c". In Region 2 these three subdivisions denote the following: subdivision "a" shows the slight increase in M_I during the closing of the bypass valve; subdivision "b" shows the nearly linear increase in M_I during the first dead-end interval where both the bypass and tank valves are closed; and subdivision "c" shows the initial increase in M_I as the tank valve just begins to open followed by its rapid decrease as the inventory volume gas is sucked into the nearly evacuated collection tank through the fully opened tank valve. The three subdivisions in Region 4 are similar to those in Region 2 and denote the following: subdivision "a" shows the slight increase in M_I as the tank valve is closing; subdivision "b" shows the increase in M_I during the second dead-end interval; and subdivision "c" shows the initial increase in M_I followed by its rapid drop off to match the atmospheric pressure condition when the bypass is fully opened.

For the lowest uncertainty, the collection time measurement should begin in the first dead-end interval (*i.e.*, Region 2b) and end in the second dead-end interval (*i.e.*, Region 4b). If the collection time began or ended in any other region, the uncertainty in mass flow could be substantially larger. For example, if the collection time began while the bypass valve was closing (Region 2a), the gas emanating from the CFV could escape into the room through the partially opened bypass valve. The uncertainty attributed to airflow leaking into or out of the bypass is difficult to quantify, and thereby increases the mass flow uncertainty.

An increase in the mass flow uncertainty also occurs if collection time begins in Region 2c while the tank valve is opening. In this case, the initial mass in the collection tank, M_T^i, must be measured dynamically (*i.e.*, while mass is accumulating in the tank) rather than statically. Figure 3 shows the increase in M_T attributed to mass flow through the partially opened tank valve in Region 2c. Since dynamic mass determinations have larger uncertainties than static determinations, it is not advantageous to begin the collection time in this region. By default,

Region 2b is the best choice to begin the collection time. Similar arguments can be made to show that Region 4b is the best choice to stop the collection time.

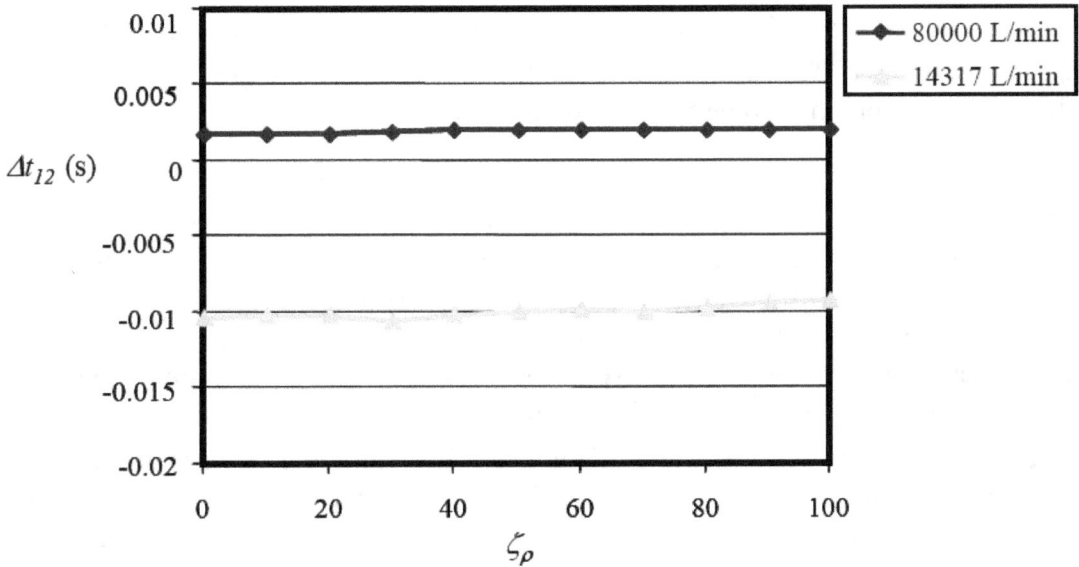

Figure 4. Time adjustment factor versus the percent density overlap parameter.

Unlike M_T, the nature of the diversion process necessitates that M_I be measured dynamically. Because M_I must be measured dynamically, the initial and final inventory mass measurements must coincide with t^i (*i.e.*, the start of collection time) and t^f (*i.e.*, the end of collection time). On the other hand, the initial and final mass measurements in collection tank do not need to coincide with t^i and t^f. For example, the initial mass of the gas in the tank, M_T^i, can be measured any time starting from the closing of the vacuum exhaust valve (at the latter part of Region 1) until just before the tank valve starts to open (at the beginning of Region 2c). Likewise, the final mass, M_T^f, can be measured any time in Region 4b, 4c, or 5. During either of these time intervals the collection tank is isolated so that the mass of gas in its interior remains constant as shown in the Fig. 3. In practice, however, the time traces of these mass measurements are not constant, but asymptote toward a constant value as the spatial pressure and temperature gradients in the gas dissipate. If there are no leaks, any non-uniformities in the time traces of the initial mass (in Regions 1, 2a, and 2b) or the final mass (in Regions 4b, 4c, and 5) are a result the method used for measuring the mass (*i.e.*, via pressure, temperature, volume, and an equation of state) and *not* an actual change in the mass. During a calibration, M_T^i and M_T^f are determined only after sufficient time is allotted to allow the gas to equilibrate as discussed in steps 3 and 7 of the operating procedures in section 4.4.

In Fig. 3 the duration of the collection time period is shown by the horizontal line that extends between Δt_1 and Δt_2. The shaded region, called the *density overlap region*, denotes all of the plausible collection times consistent with the inventory mass cancellation technique. In practice, the *percent density overlap parameter*

$$\zeta_\rho = 100\left[\frac{\rho_{match} - \rho_{min}}{\rho_{max} - \rho_{min}}\right] \qquad (10)$$

determines which of the manifold of possible collection times is used to calculate the mass flow. Here, ρ_{min} and ρ_{max} are the lower and upper limits of the density overlap region, and ρ_{match} is the matched density. From the geometry in the figure, the collection time is

$$\Delta t = \Delta \tau + \Delta t_{12} \qquad (11)$$

where the base time period, $\Delta \tau$, extends from the start of the first dead-end period to the start of the second dead-end period, and the *time adjustment factor*, $\Delta t_{12} \equiv \Delta t_2 - \Delta t_1$, is defined as the difference between the two time intervals, Δt_1 and Δt_2, where the subscripts "1" and "2" indicate which of the two dead-end periods the time intervals occur. As shown in Fig 3., these time intervals persist for a fraction of their respective dead-end intervals. Although the duration of both Δt_1 and Δt_2 depend on ζ_ρ, the time adjustment factor (Δt_{12}) should ideally have no dependence on the percent density overlap parameter. The almost uniform distribution of the time adjust factor shown in Fig. 4 confirms that it is nearly independent of ζ_ρ. In this figure the time adjust factor is measured at two flows, 14317 L/min and 80000 L/min, corresponding to collection times of approximately 109 s and 20 s respectively. The relative uncertainty of Δt_{12} due to its dependence on ζ_ρ is defined as the ratio of its standard deviation ($\sigma_{\Delta t_{12}}$) to the collection time (Δt). For longer collection times (*i.e.*, lower flows), Δt increases while $\sigma_{\Delta t_{12}}$ remains nearly fixed so that its relative uncertainty decreases. Thus, we selected two relatively large flows to determine an upper uncertainty bound. At the largest flow the relative uncertainty is $\sigma_{\Delta t_{12}}/\Delta t = 5 \times 10^{-6}$. This value is one of the contributing components for the collection time uncertainty discussed in section 5.2.2.

5. Uncertainty of *PVTt* Subsidiary Components

The mass flow determinations of a *PVTt* flow standard rely on accurate measurements of pressure, volume, temperature, and time and on the reference parameters R_u, $\mathcal{M}$, and Z. In general, the largest uncertainties in mass flow can be attributed to the measurements of volume, temperature, and pressure. However, timing measurements can also play an important role near the maximum flow capacity of a *PVTt* system when collection times are shortest. For these short collections, the largest contribution from timing uncertainties is typically associated with timing errors introduced by the flow diversion processes. On the other hand, at the lower flow capacity the collection times are longer and timing measurements typically play only a minor role in the mass flow uncertainty budget. The reference parameters, $\mathcal{M}$ and Z, are well known for common gases (*e.g.*, air, N_2, CO_2, Ar, He, etc), and contribute little to the mass flow uncertainty.

In this section, the uncertainty of the various reference parameters and measured quantities are assessed. We begin with the reference parameters, followed by the timing system, the pressure and temperature measurements in the inventory volume, the size of the inventory volume, the pressure and temperature measurements in the collection tank, and the size of collection tank. Throughout the document all of the uncertainty components are categorized as being either Type A (*i.e.*, those which are evaluated by statistical methods) or Type B (*i.e.*, those which are evaluated by other means) as described in [12]. Uncertainties having subcomponents belonging to both Type A and Type B are categorized as (A, B) as specified in [12].

5.1 Reference Parameters ($\mathcal{M}$, R_u, and Z)

5.1.1 Universal Gas Constant

The universal gas constant has a value of $R_u = 8314.472$ J/(kg·K) with a Type B relative standard uncertainty of $[u(R_u)/R_u] = 1.7 \times 10^{-6}$ [11].

Table 4. Composition of dry air.

Species	Mole Fraction (x_k)
Nitrogen	0.780849
Oxygen	0.209478
Argon	0.00934
Carbon Dioxide	0.000314
Neon	1.82×10^{-5}

5.1.2 Molecular Mass

In this work the value used for the molecular mass of dry air is $\mathcal{M} = 28.9647$ kg/kmol. It is computed using the Refprop Thermodynamic Database [10] for the composition shown in Table 4. The relative molecular mass has two sources of uncertainty: 1) a Type B uncertainty attributed to the air moisture level, and 2) a Type A uncertainty resulting from the variation in the composition of dry air. The air moisture level is maintained below 3 % relative humidity (RH). Since the RH measurement is made under room temperature conditions at a nominal pressure of $P = 800$ kPa, the mole fraction of water vapor is 9.82×10^{-5} resulting in a relative standard uncertainty attributed to air moisture level of 37×10^{-6}.

Various references list slight difference in the composition of dry air at sea level [13-15]. We estimated that the relative standard uncertainty attributed to the variation in composition is 35×10^{-6}. Thus, propagation of these two uncertainty components yields a combined relative standard uncertainty of $[u(\mathcal{M})/\mathcal{M}]_{air} = 51 \times 10^{-6}$.

5.1.3 Compressibility Factor

The compressibility factor is determined using the Refprop Thermodynamic Database [10] in conjunction with the corresponding measurements of pressure and temperature. The ranges of the pressures and temperatures in the collection tank differ from those in the inventory volume so that the uncertainties of the compressibility factors corresponding to these ranges also differ. In the collection tank the temperature ranges from 290 K to 310 K and the pressure ranges from 0 kPa to 110 kPa. For this range of conditions the relative standard uncertainty of the compressibility factor is estimated to be no more than $[u(Z_T)/Z_T] = 50 \times 10^{-6}$ [10]. In the inventory volume the temperature ranges from 290 K to 340 K and the pressure ranges from 100 kPa to 450 kPa. For this range of conditions the relative standard uncertainty of the compressibility factor is conservatively estimated to be no more than $[u(Z_I)/Z_I] = 100 \times 10^{-6}$. Both of these uncertainty components are Type B.

5.2 Collection Time

The collection time, defined previously in Eqn. (11), consist of the base time, $\Delta\tau$, and the time adjustment factor, Δt_{12}. Applying the method of propagation of uncertainty [16] the corresponding collection time uncertainty is[6]

$$\left[\frac{u(\Delta t)}{\Delta t}\right]^2 = \left(\frac{\Delta\tau}{\Delta t}\right)^2 \left[\frac{u(\Delta\tau)}{\Delta\tau}\right]^2 + \left[\frac{u(\Delta t_{12})}{\Delta t}\right]^2 \qquad (12)$$

where the relative uncertainty of the time adjustment factor (the second term) is normalized by the collection time, Δt, instead of Δt_{12}. Moreover, since $\Delta t_{12} \ll \Delta t$ (as explained in section 4.5) the ratio $\Delta\tau/\Delta t$ in the first term is close to unity. The total relative standard uncertainty for the collection time is $[u(\Delta t)/\Delta t] = 15 \times 10^{-6}$. It is comprised of the following two components: 1) the base time measurement (2×10^{-6}), and 2) the time adjustment factor (14×10^{-6}). These components are itemized in Table 5 for a 20 s collection period (*i.e.*, the shortest collection period used) and the uncertainty value of each is discussed here. The abbreviations in the table have the following meanings: Abs. Unc. is the Absolute Uncertainty, Rel. Std. Unc. is the relative standard uncertainty, Sens. Coeff. is the dimensionless sensitivity coefficient, Unc. Type is the uncertainty type, and Perc. Contrib. is the percent contribution to the combined uncertainty in collection time.

Table 5. Collection time uncertainty for a 20 s collection.

Collection Time Uncertainty.	Abs. Unc.	Rel. Std. Unc. (k=1)	Sens. Coeff.	Perc. Contrib.	Unc. Type	Comments
Collection time, Δt = 20 s	(ms)	($\times 10^{-6}$)	(-----)	(%)	(-----)	
Base time, $\Delta\tau$	0.04	2	≈1	1.9	A	Calib. of HP Counters
Time adjustment factor, Δt_{12}	0.29	14	1	98.1	B	See section 5.2.2.
Combined Uncertainty	**0.29**	**15**		**100**		

5.2.1 Base Time Measurement

The base time spans the time interval from the beginning of the first dead-end interval to the beginning of the second dead-end interval. It is measured with a redundant pair of HP counters each having a relative standard uncertainty of 2×10^{-6}. The redundancy provided by two counters helps prevent against erroneous time measurements should one of them malfunction. The HP counters are triggered by the voltage output of an electric circuit. A photodiode sensor aligned with the closed position of the bypass valve activates the electric circuit and starts the time measurement during the first flow diversion. In a similar manner, the time measurement is terminated during the second flow diversion by another photodiode sensor that produces a voltage signal when the tank valve reaches its fully closed position. Timing errors associated

---

[6] For convenience all equations symbolically expressing uncertainty are given as the variances rather than standard uncertainties unless otherwise noted.

with misalignment of the triggering signal and the valve fully closed positions of either valve are inherently accounted for by the inventory mass cancellation technique. For example, if during the first flow diversion the triggering signal is set off prematurely before the bypass valve is fully closed, the measured base time, $\Delta\tau$, will be slightly longer than its actual value. However, the measured time interval Δt_1 will be extended by the same amount so that the collection time as calculated by Eqn. (11) is invariant. Consequently, misalignment of the triggering signal does not contribute to the uncertainty. Nevertheless, proper mass accounting requires that the tank valve remain closed until the bypass valve is fully closed.

5.2.2 Time Adjustment Factor

The time adjustment factor is a small correction that adjusts the time measurement to ensure mass cancellation in the inventory volume. The time adjustment factor is evaluated by taking the difference between the time intervals Δt_1 and Δt_2. The first interval, Δt_1, begins during the first diversion period when a photodiode is activated by the closing of the bypass valve. The photodiode triggers an electric circuit that in turn outputs a voltage signal that starts the time measurement. Similarly, the measurement of Δt_2 starts during the second flow diversion when the photodiode on the tank valve is activated by its closing. The duration of both Δt_1 and Δt_2 are based on the percent density overlap parameter, ζ_ρ. In particular, measurements of pressure and temperature are used to calculate the density time histories during the 1st and 2nd dead-end intervals, and ζ_ρ selects the particular matched density from the region of density overlap. Since the voltage, pressure, and temperature measurements used to determine the duration of Δt_1 and Δt_2 are acquired by a data acquisition card sampling at 3000 Hz, the resolution of the calculated time intervals is limited to 0.33 ms. If a rectangular distribution is assumed, the standard uncertainties for both Δt_1 and Δt_2 equal 0.19 ms, so that for a 20 s collection the corresponding relative uncertainties are 10×10^{-6}.

The total uncertainty in Δt_{12} consists of three components. These include the uncertainty attributed to Δt_1 (10×10^{-6}), the uncertainty attributed to Δt_2 (10×10^{-6}), and the uncertainty attributed to the uniformity of Δt_{12} with ζ_ρ (5×10^{-6}) discussed previously in section 4.5. The first two are Type B uncertainties while the third is a Type A uncertainty. Propagation of these three components yields a total relative standard uncertainty for the time adjustment factor equal to 14×10^{-6}.

5.3 Pressure and Temperature in the Inventory Volume

The initial pressure measurement in the inventory volume is obtained by averaging the results of two fast pressure transducers during the first flow diversion. The final pressure is measured by averaging the readings of the same two transducers during the second flow diversion. The first transducer is positioned adjacent to the bypass valve and the second is located next to the tank inlet valve. The initial and final temperatures are determined by averaging the results of two type T thermocouples of 0.025 mm nominal diameter. The thermocouples are positioned adjacent to the pressure sensors, one next to the tank valve and the other next to the bypass valve.

Figure 5 shows the time histories for the pressure (left) and the temperature (right) during the first and second flow diversions for a nominal flow of 0.4 kg/s. This data is acquired using a data acquisition card sampling at 3000 Hz. The beginning of both the first and second dead–end intervals starts at $t = 0$ s. The pressure and temperature time traces begin at near ambient conditions, increase as mass accumulates into the inventory volume, and then sharply decrease as the accumulated mass is exhausted either into the nearly evacuated collection tank (*i.e.*, 1st flow diversion) or to the bypass at ambient conditions (*i.e.*, 2nd flow diversion).

To capture the rapidly changing conditions in the inventory volume during flow diversions, both the pressure and temperature sensors must have a fast time response. The reading indicated by a slow sensor will lag behind the actual value. The error associated with a slow sensor can be predicted if the transducer time constant is known. The typical manufacturer specified time constant for the pressure transducer is $\tau_P = 3$ ms. The thermocouple time constant depends on flow. In a previous work, the thermocouple time constant was measured to be 20 ms at a flow of 1 g/s [17]. This value agreed to within 70 % of the theoretical value that was predicted using an empirical heat transfer coefficient corresponding to flow over a small diameter cylinder [18]. No attempt was made to obtain better agreement between the measured and predicted time constant since the net effect of the sensor time response has little impact on the mass flow uncertainty. In fact, we assumed that the thermocouple time constant remained fixed at 20 ms, instead of decreasing at larger flows as indicated by experimental and theoretical evidence [17, 18]. This assumption is also justified by the small impact of this parameter on flow uncertainty.

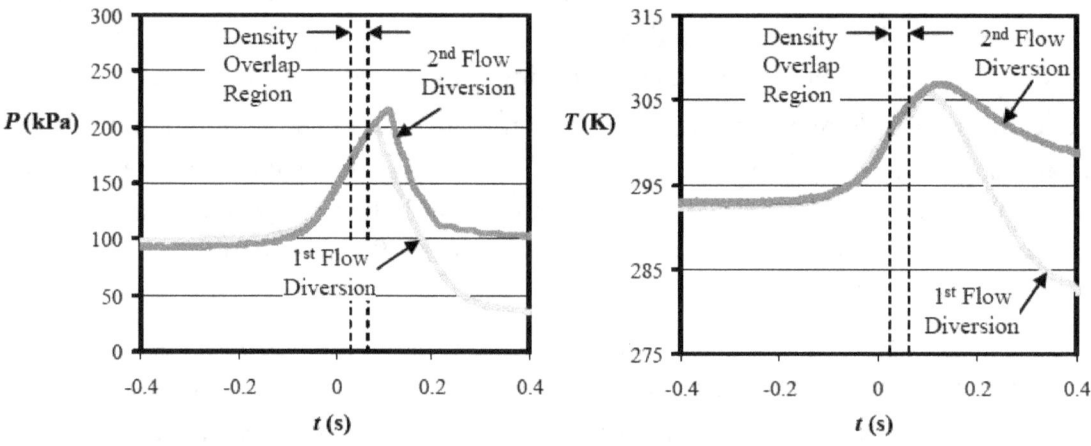

Figure 5. Time histories of the inventory volume pressure (left) and temperature (right) during the first and second flow diversions for a nominal flow of 0.4 kg/s. (Data collected using fast pressure transducers and type T thermocouples.)

Figure 5 shows the complete pressure and temperature time histories during diversion processes. However, only a small fraction of this time history is critical for computing mass flow. The inventory mass cancellation technique only requires the pressure and temperature data occurring within the density overlap region (see section 4.5). Given that the pressure and temperature time traces in this region are almost identical (*i.e.*, a symmetric diversion process), and that the same transducers are used to make both pairs of measurements, several of the sources of uncertainty are correlated. Moreover, the correlated quantities cancel almost completely when the inventory mass cancellation technique is implemented as part of the flow calibration process. If the diversion process was asymmetric, these correlated uncertainties would not cancel, and the corresponding uncertainties from these components could increase significantly.

Below we assess the uncertainty for pressure and temperature measurements in the inventory volume. Since the greatest inventory volume uncertainties occur at the largest flows, the analysis gives the uncertainties at the largest flow (77000 L/min).

5.3.1 Initial Pressure in the Inventory Volume

The initial pressure uncertainty components are itemized in Table 6. These six components include 1) the calibration fit residuals, 2) the transducer mounting orientation 3) the response time of the sensor, 4) the spatial sampling error 5) the ambient temperature effect, and 6) the sensor repeatability. The first three sources of uncertainty are perfectly correlated since neither their sign nor magnitude change between the initial and final measurement. The remaining three sources of uncertainty are treated as uncorrelated. Propagation of the uncorrelated sources yields a total relative standard uncertainty of $[u_u(P_I^i)/P_I^i] = 2.3\%$, while propagation of the correlated sources gives $[u_c(P_I^i)/P_I^i] = 4.0\%$. The total relative standard uncertainty is obtained by propagating the uncorrelated and correlated sources, thereby yielding $[u(P_I^i)/P_I^i] = 4.7\%$. An evaluation for each of these uncertainty components is provided below, beginning with the uncorrelated sources and followed by the correlated sources.

Table 6. Uncertainty of the initial pressure measurement in the inventory volume.

Uncertainty of initial inventory pressure	Abs. Unc.	Rel. Std. Unc. (k=1)	Perc. Contrib.	Unc. Type	Comments
Initial Pressure, $P_I^i = 190.9$ kPa	(kPa)	(%)	(%)	A or B	
Uncorrelated Unc.					
Spatial sampling error	4.41	2.3	24.6	B	Meas. pres. Difference
Ambient temperature effects	0.12	0.1	0.0	B	Manuf. spec.
Sensor repeatability	0.17	0.1	0.0	B	Manuf. spec.
Correlated Unc.					
Calibration fit residuals	0.75	0.4	0.7	A	End-to-end calibration to Paros 1400 kPa
Time response of Heise transducer	7.68	4.0	74.7	B	Dead-End Flow Model
Transducer mounting orientation	0.0	0.0	0.0	A	Always in same mounting position
Propagation of Uncorrelated Sources	4.41	2.3	24.6		
Propagation of Correlated Sources	7.72	4.0	75.4		
Combined Uncertainty	8.89	4.7	100		

Among the three uncorrelated uncertainty components, the uncertainty attributed to spatial sampling errors is by far the largest. We determined this uncertainty experimentally while the other two uncertainty components (the ambient temperature effect and the sensor repeatability), were obtained via manufacturer specifications. Based on these specifications both of these components have relative standard uncertainties equal to 0.1 %.

The sampling error is defined as the difference between the calculated average pressure (from the two Heise transducers) and the actual average pressure in the inventory volume. Sampling errors are caused by pressure gradients formed within the inventory volume during the dead-ended intervals. These pressure gradients are caused by two sources: 1) by the low-pressure jet exhausting from the CFV stagnating against the closed tank and bypass valves, and 2) by the pressure impulse attributed to closing either the bypass valve (*i.e.*, 1st dead-end interval) or the tank valve (*i.e.*, 2nd dead-end interval) just prior to the start of the dead-end periods. Because the CFV mass flow, and the initial inventory pressures, and temperatures are nearly the same during the first and second dead-end intervals, the size and location of pressure gradients formed during these periods are expected to be similar and to some extent correlated. However, no attempt was made to assess the degree of correlation between the initial and final pressure fields. Instead, we conservatively treated the spatial sampling error as an uncorrelated uncertainty component. To this end, the initial and final spatial sampling errors are evaluated independently by two separate experiments. In each experiment we measured the pressure at the locations in the inventory volume where the largest pressure differences are expected. Pressure measurements are made at the exhaust of the CFV where we expect the lowest pressure, and adjacent to the bypass and tank inlet valves where the flow stagnates and the largest pressures are expected. At the maximum flow, the largest pressure difference between these locations is only 2.3 % of the initial average pressure, and the sampling error is defined equal to this pressure difference.

The correlated uncertainties include the calibration fit residuals, transducer orientation, and the sensor response time. Experimental records show the relative standard uncertainty of the calibration fit residuals is 0.4 %. There is no uncertainty attributed to transducer orientation since the sensors are calibrated and used in the same orientation. The time response of the sensor is estimated using a semi-empirical mathematical model. The model calculates the pressure increase during the dead-end interval assuming the process is isentropic. The isentropic pressure response is linearized over the dead-end period and used with the sensor time constant in a first order differential model to predict the pressure lag. This model is a simplified version of a more complex model given in [17]. Although this model is not as accurate, it gives reasonable results that are appropriate for the relatively minor importance of this uncertainty component. The predicted pressure lag of this simplified model is

$$\Delta P_{lag} = P_{atm}\left(\frac{\tau_P}{\Gamma_{DE}}\right)\left[\left(\frac{\dot{m}\Gamma_{DE}}{\rho_{atm}V_I}+1\right)^{\gamma} - 1\right] \tag{13}$$

where τ_P is the time constant of the Heise transducer, Γ_{DE} is the *effective* dead-end interval (*i.e.*, the actual dead-end period plus half the time required to close either the tank or bypass valve), $\gamma = 1.4$ is the specific heat ratio for air, $P_{atm} \approx 101.325$ kPa is the initial atmospheric pressure in the inventory volume just before the start of the diversion process, $\rho_{atm} \approx 1.2$ kg/m^3 is the initial density under ambient conditions, and V_I is the size of the inventory volume. From Eqn. (13), the pressure lag increases with increasing mass flow, longer dead-end intervals, and smaller inventory volume sizes. Experience indicates that the mass flow has the most significant effect since it varies significantly over the operating range of the *PVTt* flow standard. For example, at the largest flow (77000 L/min), the predicted pressure lag is 7.68 kPa, but makes only a negligible contribution at the lower flows (7000 L/min or below).

5.3.2 Final Pressure in the Inventory Volume
Both the initial and final pressures are measured with the same transducers. Consequently, the final pressure uncertainty has the same six uncertainty components as the initial pressure. Moreover, each of the six uncertainty components has same uncertainty type (*i.e.*, Type A or B) as shown previously in Table 6. While the absolute values of uncertainty for these six components is the same for both the initial and final pressure measurements, the relative values

can differ slightly attributed to differences between the initial and final pressures. The relative standard uncertainty of the correlated components of the final pressure include the calibration fit residuals (0.4 %), the transducer mounting orientation (0 %), and the sensor response time (4.1 %). The uncorrelated components include the spatial sampling error (2.3 %), the ambient temperature effect (0.1 %), and the sensor repeatability (0.1 %). The total uncorrelated uncertainty is $[u_u(P_I^f)/P_I^f] = 4.1\%$ while the total correlated uncertainty is $[u_c(P_I^f)/P_I^f] = 2.3\%$ so that the total uncertainty is $[u(P_I^f)/P_I^f] = 4.7\%$.

5.3.3 Initial Temperature in the Inventory Volume

The four uncertainty components for the initial temperature are categorized into uncorrelated and correlated components and are shown in Table 7. The uncorrelated components include the spatial sampling error and the thermocouple repeatability while the correlated uncertainties include the sensor time response and the correction for the moving fluid stagnating against the thermocouple surface (*i.e.*, static versus stagnation). The total uncorrelated uncertainty is $[u_u(T_I^i)] = 6.0$ K and the total correlated uncertainty is $[u_c(T_I^i)] = 33.6$ K. The correlated and uncorrelated components are propagated to give a total relative standard uncertainty of $[u(T_I^i)] = 34.1$ K.

Table 7. Uncertainty of the initial inventory temperature measurement.

Uncertainty of initial inventory temperature	Abs. Unc. (k=1)	Rel. Std. Unc.	Perc. Contrib.	Unc. Type	Comments
Initial Temperature, $T_I^i = 345.2$ K	(K)	(%)	(%)	(A or B)	
Uncorrelated Unc.					
Temperature spatial sampling error	6.0	1.7	3.1	A	Meas. temp. difference
Repeatability	0.2	0.1	0.0	B	Manuf. Spec. of thermistor used for cold junction compensation
Correlated Unc.					
Thermocouple time response	33.3	9.5	95.0	B	Dead-End Flow Model
Static vs. stagnation	4.7	1.4	1.9	B	See section 5.3.3
Propagation of Uncorrelated Sources	**6.0**	**1.7**	**3.1**		
Propagation of Correlated Sources	**33.6**	**9.6**	**96.9**		
Combined Uncertainty	**34.1**	**9.7**	**100**		

The spatial sampling error is determined experimentally by measuring the temperatures adjacent to the bypass and tank valves at the maximum flow (77000 L/min). The largest measured temperature difference is less than 6.0 K. The repeatability of the sensor is conservatively estimated to be 0.2 K, and the correction for the static temperature versus measured temperature is 4.7 K. The uncertainty attributed to the sensor time response is 33.3 K. It is calculated using

$$\Delta T_{lag} = T_{atm}\left(\frac{\tau_T}{\Gamma_{DE}}\right)\left[\left(\frac{\dot{m}\Gamma_{DE}}{\rho_{atm}V_I}+1\right)^{\gamma-1}-1\right] \tag{14}$$

where τ_T is the temperature time constant of the thermocouple, and $T_{atm} \approx 293.15$ K is the initial ambient temperature in the inventory volume. This expression is based on the isentropic model discussed previously in section 5.3.1.

5.3.4 Final Temperature in the Inventory Volume

The final temperature has the same uncertainty components as the initial temperature and the corresponding uncertainties types are the same. These include the spatial sampling error (6.0 K), the sensor repeatability (0.2 K), the sensor time response (34.0 K), and the dynamic correction for the static temperature measurement (4.7 K). The total uncorrelated uncertainty is $[u_u(T_I^f)] = 6.0$ K, and the total correlated uncertainty is $[u_c(T_I^f)] = 34.3$ K. The correlated and uncorrelated components are propagated to give a total relative standard uncertainty of $[u(T_I^f)] = 34.8$ K.

5.4 Inventory Volume

The size of the inventory volume is adjusted as necessary to accommodate the quantity of flow. Larger flows require larger inventory volumes to prevent the pressure rise during the dead-ended periods from unchoking the CFV. If the CFV unchokes, then the corresponding decrease in mass flow violates the steady state assumption used in deriving Eqn. (8), and thereby introduces additional uncertainty. Fortunately, the uncertainty in the size of the inventory volume does not play a significant role in uncertainty analysis. The inventory mass cancellation technique causes its corresponding sensitivity coefficient to be zero (see section 6.2), and consequently, the uncertainty attributed to the size of the inventory volume is also zero. The size of the inventory volume does however have a small effect on the overall mass flow uncertainty through its influence on the inventory pressure and temperature sensitivity coefficients. As such, reasonably accurate values must be used. We measure the size of inventory volume to within 25 % of its actual size using a tape measure as discussed previously in section 4.5. Since its value is obtained using only a single measurement it is a Type B uncertainty.

Table 8. Uncertainty of the initial tank pressure.

Uncertainty of Initial Tank Pressure	Abs. Unc. (k=1)	Rel. Std. Unc.	Perc. Contrib.	Unc. Type	Comments
Initial Tank Pressure, $P_T^i = 0.1$ kPa	(Pa)	($\times 10^{-6}$)	(%)	(-----)	
Transducer Accuracy	0.125	1250	4.2	B	Manuf. Spec.
Ambient Temperature Effect	0.160	1600	6.8	B	Manuf. Spec. (0.04 % reading/°C from 22°C)
Drift from Cal. Records	0.557	5774	89.0	A	<1 Pa per year, assume rect.
Spatial Gradients in Pressure	0.001	14	0.0	B	Based on Hydrostatic Pressure Head
Combined Uncertainty	**0.612**	**6120**	**100**		

5.5 Pressure and Temperature in the Collection Tank

5.5.1 Initial Tank Pressure

The initial tank pressure is measured by averaging the result of two 1333.22 Pa (10 Torr) MKS capacitance diaphragm gages, each with a manufacturer specified relative uncertainty of 0.25 % taken to be at the 95 % confidence level. Additional uncertainties are attributed to ambient temperature effects, to zero drift, and to spatial pressure gradients in the tank. All of these uncertainty components are shown in Table 8. The total relative standard uncertainty attributed to the initial pressure measurement is $[u(P_T^i)/P_T^i] = 6120 \times 10^{-6}$.

5.5.2 Final Tank Pressure

The final pressure in the collection tank is measured using a Paroscientific Model pressure transducer with a full scale of 200 kPa. This transducer is calibrated at six month intervals using a Ruska piston pressure gauge whose piston area is traceable to the NIST Pressure and Vacuum Group [6]. The relevant uncertainty components for pressure are itemized in Table 9 including the calibration of the pressure transducer, (17×10^{-6}); the measured drift limit from calibration records, (60×10^{-6}); the calibration fit residuals, hysteresis, and thermal effects, (100×10^{-6}), and spatial gradients in the tank attributed to the hydrostatic pressure head (0.5×10^{-6}). The propagation of these components yields a total relative pressure uncertainty of $[u(P_T^f)/P_T^f] = 118 \times 10^{-6}$.

Table 9. Uncertainty of the final tank pressure.

Uncertainty of Final Tank Pressure	Abs. Unc.	Rel. Std. Unc. (k=1)	Perc. Contrib.	Unc. Type	Comments
Final Tank Pressure, $P_T^f = 95.25$ kPa	(Pa)	($\times 10^{-6}$)	(%)	(A or B)	
Transfer standard for static pres.	1.6	17	2.1	B	Ruska Piston Pres. Gauge
Drift from Cal. Records	5.7	60	25.9	A	< 0.01 % in 6 months, assume rect.
Residual, hystersis, thermal effects	9.5	100	72.0	A	From cal. records expts.
Spatial pressure gradients in Tank	0.05	0.5	0.0	B	Based on hydrostatic pressure head
Combined Uncertainty	**11.2**	**118**	**100**		

5.5.3 Initial and Final Average Gas Temperature in the Collection Tank

Both the initial and final gas temperatures are measured by averaging 37 thermistors distributed throughout the collection tank. Because the collection tank is initially evacuated, the sensitivity coefficient corresponding to the initial temperature is significantly lower than the final temperature. As a result the initial temperature measurement only requires marginal accuracy relative to the final temperature measurement. Therefore, significantly more effort is spent obtaining a low uncertainty final temperature measurement. In this analysis the standard uncertainty of the initial temperature measurement is $[u(T_T^i)] = 1206$ mK while the standard uncertainty for the final temperature is $[u(T_T^f)] = 64.6$ mK. The various uncertainty components comprising the initial and final temperature measurements are evaluated below.

The standard uncertainty components for the final temperature measurement are shown in Table 10. These components include the thermistor calibration transfer standard (1.2 mK), the uniformity of the temperature bath used for calibrations (1.0 mK), the standard deviation of the calibration fit residuals (7 mK), the manufacturer specified drift (28.9 mK), radiation and self-heating (1.8 mK), the thermistor time response (2.5 mK), and the spatial sampling error (57.3 mK). The most significant of these uncertainties are the spatial sampling error and the thermistor drift, which together contribute almost 99 % of the uncertainty. The uncertainty attributed to thermistor drift (28.9 mK) is obtained by dividing the manufacturer specified drift limit of 50 mK (taken to be at the ninety-five percent confidence level) by $\sqrt{3}$ as prescribed for a rectangular distribution. The uncertainty attributed to drift can be decreased, if necessary, by calibrating the thermistors more frequently. The 50 mK drift limit is based on a five year calibration schedule (*i.e.*, the drift rate is 10 mK/year). The five year interval was selected to avoid difficulties associated with retrieving the thermistors inside the collection tank. We may, in the future, select to calibrate the thermistors at shorter intervals to further reduce the uncertainty.

Table 10. Uncertainty of the final tank temperature.

Uncertainty of Final Tank Temperature Final tank temperature, T_T^f = 294 K	Abs. Unc. (mK)	Rel. Std. Unc. (k=1) ($\times 10^{-6}$)	Perc. Contrib. (%)	Unc. Type (A or B)	Comments
Temperature transfer standard	1.2	4	0.0	B	Traceable to NIST temperature group
Uniformity of temperature bath	1.0	3	0.0	B	Expt. varied position of Temp. Std.
Fit residuals	7.0	24	1.2	A	Based on calibration data
Drift (I, R, DMM, thermistors)	28.9	98	20.0	B	Manuf. spec < 50 mK/5 year, assume rect distribution.
Radiation, self-heating	1.8	6	0.1	A	Expt. varied current & calculated
Thermistor time response	2.5	8	0.1	B	Est. based on theoretical model
Temperature spatial sampling error	57.3	195	78.6	A	Expt. measured [1]
Combined Uncertainty	**64.6**	**220**	**100**		

The spatial sampling error was determined experimentally by characterizing the size and decay rate of the temperature gradients in the gas after filling. Based on these measurements, a mathematical model was developed both to estimate the settling time necessary for the gas in the tank to thermally equilibrate and to determine the best arrangement of sensors in the collection tank [19]. When the ducted fan (see Fig. 1) is used to mix the gas, the settling time for thermal equilibrium is 2700 s. The final average gas temperature is calculated after this period using a volume weighted integration of all 37 thermistors. The spatial sampling error is taken to be the root-sum-square of the volume weighted temperature differences in three locations of the collection tank including 1) temperature differences near the fan motor, 2) temperature differences in the thermal boundary layer adjacent to the tank wall, and 3) temperature differences in far-field.

The spatial sampling error has been reduced from 70.3 mK as given in a previous publication [19] to its current value of 57.3 mK. In the original calculation of the spatial sampling error, six months of temperature measurements indicated that the temperature difference in the region close to the fan motor was 500 mK, in the boundary layer region the temperature difference was 250 mK, and in the far-field it was 62 mK. Given that the region near the fan motor accounts for 0.6 % of the collection tank volume, the boundary layer region accounts for 20 %, and the far-field accounts for the remaining 79.4 %, the volume weighted temperature differences are 3 mK, 50 mK, and 49.4 mK, respectively. A root-sum-square of these three components gave a sampling error of 70.3 mK. However, additional temperature measurements over the past three years have demonstrated that the uncertainty attributed to temperature differences in the boundary layer region can be reduced from 50 mK (as given in [19]) to its present value of 28.9 mK.

The original temperature characterization indicated that the temperature measurements in the boundary layer were sensitive to the degree of stratification in the room enclosing the *PVTt* flow standard.[7] Since the degree of stratification changed seasonally (*i.e.*, from the winter to summer months), we used the worst case (*i.e.*, 50 mK) as the temperature difference in the boundary layer. We conservatively used the largest temperature difference to avoid underestimating the sampling error, if for example future measurements showed a larger boundary layer effect due to seasonal

---

[7] Stratification had little impact on the temperature differences in the far-field or near the fan motor.

temperature changes. However, additional temperature measurements suggest that 50 mK is a reasonable upper bound. Moreover, an array of 14 thermocouples distributed along the outer surface of the collection tank verifies for each calibration that temperature stratification never exceeds the worse case. Since 50 mK is the maximum possible value, we assume a rectangular distribution so that this value is divided by $\sqrt{3}$ and the standard boundary layer temperature difference is 28.9 mK.

Several of the uncertainty components for the initial temperature measurement are identical to those of the final temperature measurement. These include the thermistor calibration transfer standard (1.2 mK), the uniformity of the temperature bath used for calibrations (1.0 mK), the calibration fit residuals (7 mK), and the manufacturer specified drift (28.9 mK). The remaining uncertainty components, including the thermistor time response (303 mK), the radiation and self-heating (167 mK), and the spatial sampling error (1155 mK), are all significantly larger than the final temperature measurement. These components are larger because the heat transfer mechanisms that affect the temperature measurements are drastically different between the initial and final conditions in the collection tank. In the final condition, the collection tank is at near atmospheric conditions and the heat generated by joule heating in the thermistor is dissipated mainly by natural convection processes. On the other hand, the tank is initially under vacuum conditions so that radiation is the only significant mode of heat transfer from the thermistor. The radiation heat transfer dissipates heat less effectively, and has a much lower heat transfer coefficient. The lower heat transfer coefficient results in the slower sensor time response and higher value of self-heating. In a similar manner, the poor mixing under vacuum conditions results in the larger spatial sampling error.

5.6 Collection Tank Volume

The collection tank volume is determined using a gravimetric weighing procedure whereby a measured mass of gas is transferred into the initially evacuated collection tank. The volume of the tank is determined by dividing the mass of gas transferred into the tank by the change in gas density attributed to filling. When the gas that remains trapped in the volume of tubing connecting the supply gas to the tank is considered, the tank volume is

$$V_T = \frac{\Delta M_{cyl}}{\rho_T^f - \rho_T^i} - V_c \qquad (15)$$

where ΔM_{cyl} is the mass of gas transferred to the tank, ρ_T^i is the initial density measurement in the tank, ρ_T^f is the final density measurement in the tank, and V_c is the connecting volume between the gas source and the collection tank.

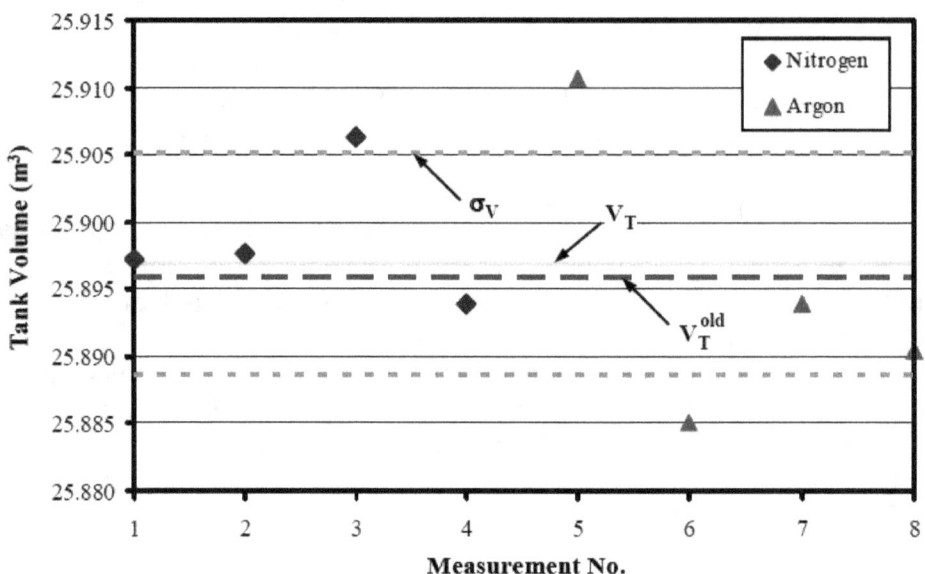

Figure 6. Shows eight measurements (four with argon and four with nitrogen) based on gravimetric weighing technique used to determine the collection tank internal volume (V_T), and the standard deviation of repeated measurements (σ_V).

The source of gas used for the volume determination was an array of high pressure gas cylinders. The mass of gas displaced into the collection tank was determined by subtracting the initial cylinder mass (before filling the tank) with the mass after the filling process. Both the initial and final masses are determined using a 600 kg Mettler IDS scale with a resolution of 0.0002 kg. The array of cylinders was connected to the tank by nylon tubing of diameter 6.35 mm (0.25 in) that served as the connecting volume. Before beginning the experiment, the collection tank was purged by repeatedly evacuating its contents and filling it with the source gas (*i.e.*, either argon or nitrogen). The experiment began by determining the initial density of the gas in the nearly evacuated collection tank via temperature and pressure measurements. Following this, the collection tank is filled with the source gas until it reaches atmospheric pressure. The volume of the collection tank is determined at atmospheric pressure to match the condition that it is used during calibration, and to prevent leakage into or out of the collection tank while waiting for the gas to thermally equilibrate. Once equilibrium conditions are reached, the final gas density is determined via pressure and temperature measurements. Both the initial and final pressure and temperature measurements use the same instrumentation used during an actual calibration cycle (see section 5.5).

As shown in Fig. 6, eight independent measurements were used to determine the collection tank volume. Four of the volume determinations used nitrogen as the source gas while the remaining four used argon. The tank volume, $V_T = 25.8969$ m^3, is the average of these eight measurements. The figure shows that this value for the tank volume compares well with the previously used value, differing by only 0.0036 %. This level of agreement is not unexpected since the tank volume has no reason to change. The standard deviation of repeated measurements is indicated by the two dashed lines ($\sigma_V/V_T = 318 \times 10^{-6}$). The relative standard uncertainty attributed to repeated volume measurements equals to the standard deviation of the mean, $\frac{\sigma_V/V_T}{\sqrt{N}} = 113 \times 10^{-6}$, where $N = 8$ are the eight repeated volume measurements.

Table 11. Uncertainty of the collection tank volume.

Tank Volume Uncertainty	Rel. Std. Unc. (k=1)	Sen. Coeff.	Perc. Contrib.	Unc. Type	Comments
Tank Volume, $V_T = 25.8969$ m^3	($\times 10^{-6}$)	(-----)	(%)	(A or B)	
Connecting volume, $V_c = 128.4$ cm^3	58940	-5.0×10^{-6}	0.0	B	See Section 5.6.1
Initial tank density, $\rho_T^i = 8.55 \times 10^{-4}$ kg/m^3	8988	7.4×10^{-4}	0.1	A, B	See Section 5.6.2
Final tank density, $\rho_T^f = 1.1620$ kg/m^3	250	-1.0	82.3	A, B	See Section 5.6.3
Effects of Leaks	0	1.7×10^{-5}	0.0	B	See Section 5.6.4
Change in cylinder mass, $\Delta M_{cyl} = 30.0710$ kg	28	1.0	1.0	A, B	See Section 5.6.5
Std. dev. of the mean for the eight repeated volume meas., $\sigma_V/\sqrt{8}\,V_T$	113	1.0	16.6	A	Four volume determinations in Ar and four in N$_2$
Combined Uncertainty	**276**		**100**		

The expression for the relative standard uncertainty of the collection tank volume is determined by applying the law of propagation of uncertainty to Eqn. (15). When the uncertainty contribution from repeated measurements is included the resulting expression is

$$\left[\frac{u(V_T)}{V_T}\right]^2 = \left[1+\frac{V_c}{V_T}\right]^2\left[\frac{u(\Delta M_{cyl})}{\Delta M_{cyl}}\right]^2 + \left[\left(\frac{-1}{1-\rho_T^i/\rho_T^f}\right)\left(1+\frac{V_c}{V_T}\right)\right]^2\left[\frac{u(\rho_T^f)}{\rho_T^f}\right]^2$$
$$+ \left[\left(\frac{\rho_T^i/\rho_T^f}{1-\rho_T^i/\rho_T^f}\right)\left(1+\frac{V_c}{V_T}\right)\right]^2\left[\frac{u(\rho_T^i)}{\rho_T^i}\right]^2 + \left[\frac{V_c}{V_T}\right]^2\left[\frac{u(V_c)}{V_c}\right]^2 + \left(\frac{\sigma_{V_T}}{8V_T}\right)^2 \quad (16)$$

where the correlation between the initial and final densities, ρ_T^i and ρ_T^f, due to their common dependence on the reference parameters $\mathcal{M}$ and R_u is negligible and has been ignored. Additionally, the correlation between ρ_T^i and ρ_T^f attributed to calibration of the temperature sensors is negligible.[8] The various uncertainty components and sensitivity coefficients are listed in Table 11. When Eqn. (16) is used to combine these uncertainty components the total relative standard uncertainty attributed to the collection tank volume is $[u(V_T)/V_T] = 276 \times 10^{-6}$. The documentation for each of the components contributing to the overall uncertainty is discussed here.

5.6.1 Connecting Volume

The volume of the nylon tube used to connect the high pressure gas cylinders to the collection tank was geometrically determined by measuring its internal diameter and length. Using this

---

[8] The initial and final pressures are measured with different transducers so that there is no correlation attributed to their calibration.

method, the size of the connecting volume was 128.4 cm³ and its relative standard uncertainty is $[u(V_c)/V_c] = 58940 \times 10^{-6}$.

5.6.2 Initial Gas Density in the Collection Tank

The initial tank density was determined by using the measured pressure and temperature in conjunction with the equation of state for the appropriate gas (*i.e.*, nitrogen or argon). The expression of uncertainty for the initial density is

$$\left[\frac{u(\rho_T^i)}{\rho_T^i}\right]^2 = \left[\frac{u(P_T^i)}{P_T^i}\right]^2 + \left[\frac{u(T_T^i)}{T_T^i}\right]^2 + \left[\frac{u(Z_T^i)}{Z_T^i}\right]^2 + \left[\frac{u(R_u)}{R_u}\right]^2 + \left[\frac{u(\mathcal{M})}{\mathcal{M}}\right]^2 \qquad (17)$$

where the relative standard uncertainties for the universal gas constant (1.7×10^{-6}) has been documented in previously in section 5.1.1, and the absolute temperature uncertainty (1206 mK) has been documented in section 5.5.3. The relative standard pressure (7998×10^{-6}) is slightly larger than the value given in section 5.5.1 because a lower initial tank pressure was used for the volume determination. The relative standard uncertainties attributed to the molecular mass and compressibility factor are (11×10^{-6}) and (10×10^{-6}) respectively. The molecular mass uncertainty is attributed to gas impurities in the source gas, and the compressibility uncertainty is primarily due to the Refprop thermodynamic database used to evaluate the equation of state [6, 20]. The universal gas constant, initial gas pressure in the collection tank, molecular mass, and the compressibility factor are all Type B uncertainties. The temperature uncertainty, however, has uncertainty subcomponents of both Type A and B. The total relative standard uncertainty of the initial density measurement is $[u(\rho_T^i)/\rho_T^i] = 8988 \times 10^{-6}$. Since it is comprised of uncertainty subcomponents of both Type A and B, it is categorized as being a Type A, B uncertainty.

5.6.3 Final Gas Density in the Collection Tank

The uncertainty of the final gas density is determined using an expression analogous to Eqn. (17) with the corresponding uncertainties being of the same type. While the uncertainties in the universal gas constant, compressibility factor, and molecular weight are identical to the values used for the initial density, the uncertainties for the pressure and temperature measurements differ. The pressure and temperature uncertainties correspond to the uncertainties discussed previously in sections 5.5.2 and 5.5.3 for the final condition in the collection tank. The relative standard uncertainty in pressure is 118×10^{-6} while the standard absolute temperature uncertainty is 64.6 mK. The total relative standard uncertainty for the final gas density is $[u(\rho_T^f)/\rho_T^f] = 250 \times 10^{-6}$, and its uncertainty type is A, B.

5.6.4 Effect of Leaks on Collection Tank Volume Determination

During the volume determination, every precaution was taken to minimize the influence of leaks. The high pressure gas cylinders were checked for leaks before weighing using a soap solution. If no leaks were discovered, the array of cylinders was wiped dry and then allowed to sit over night to allow any remaining soap solution to evaporate. The next morning the initial mass of the cylinders was determined by averaging at least three separate weighings. If each of the three successive weighings decreased in value, additional weighings were performed to ensure that the cylinders were not leaking gas.

Before beginning the mass transfer into the collection volume, the connecting volume of nylon tubing was checked for leaks using a soap solution. If no leaks were found, the high pressure cylinders were emptied into the collection tank. After filling the tank, both the tank and the cylinders were near ambient pressure so that any leakage due to a pressure difference was minimal. Because of these precautions, leaks are not expected to make a meaningful contrinbution to the uncertainty of the collection tank volume.

5.6.5 Mass Transferred into the Collection Tank

The mass of gas transferred from the high pressure cylinders into the collection tank is determined by weighing the cylinders before and after the collection tank is filled. A double substitution method is used to determine the initial and final mass of the cylinders. Based on this method, the expression for either the initial or final cylinder mass is

$$M_{cyl} = M_{ref}\left(1 - \frac{\rho_{air}}{\rho_{ref}}\right) + \rho_{air} V_{cyl} + O_{DS} M_{sen}\left(1 - \frac{\rho_{air}}{\rho_{sen}}\right) \quad (18)$$

where M_{ref} is the total mass of the set of NIST traceable reference masses, M_{sen} is one of the reference masses that is used as the sensitivity mass, and O_{DS} is the correction factor corresponding to the double substitution method

$$O_{DS} \equiv \frac{(O_{cyl} + O_{cyl+sen}) - (O_{ref} + O_{ref+sen})}{2(O_{ref+sen} - O_{sen})} \quad (19)$$

The correction factor consists of the four mass observations made using the 600 kg Mettler IDS scale. These four mass observations include the following: 1) O_{cyl}, the observed mass of the cylinder by itself, 2) $O_{cyl+sen}$, the observed mass of the cylinder and sensitivity weight together, 3) $O_{ref+sen}$, the observed mass of the reference mass and sensitivity weight together, and 4) O_{ref} the observed mass of the reference mass by itself. The remaining variables in Eqn. (18) account for air buoyancy forces. These variables include the density of the room air (ρ_{air}); the densities of the stainless steel reference and sensitivity masses (ρ_{ref} and ρ_{sen} respectively); and the external volume of the gas cylinders (V_{cyl}).

The difference between the initial and final cylinder masses equals the amount of gas displaced from the cylinders into the collection tank. When the initial and final masses are computed using Eqn. (18), the gas displaced from the cylinders into the collection tank is

$$\Delta M_{cyl} = \left[M_{ref}^i + O_{DS}^i M_{sen}\right]\left(1 - \frac{\rho_{air}^i}{\rho_{ref}}\right) - \left[M_{ref}^f + O_{DS}^f M_{sen}\right]\left(1 - \frac{\rho_{air}^f}{\rho_{ref}}\right) \quad (20)$$
$$+ V_{cyl}\left(\rho_{air}^i - \rho_{air}^f\right)$$

where for these measurements the densities of the stainless steel reference and sensitivity masses are equal (i.e., $\rho_{ref} = \rho_{sen}$). The uncertainty expression for the mass of gas transferred to the collection tank is

$$\left[\frac{u(\Delta M_{cyl})}{\Delta M_{cyl}}\right]^2 = \left[\left(\frac{M_{ref}^i + O_{DS}^i M_{sen}^i}{\Delta M_{cyl}}\right)\left(\frac{\rho_{air}^i}{\rho_{ref}}\right) - \frac{V_{cyl}\rho_{air}^i}{\Delta M_{cyl}}\right]^2 \left[\frac{u(\rho_{air}^i)}{\rho_{air}^i}\right]^2$$
$$+ \left[\left(\frac{M_{ref}^f + O_{DS}^f M_{sen}^f}{\Delta M_{cyl}}\right)\left(\frac{\rho_{air}^f}{\rho_{ref}}\right) - \frac{V_{cyl}\rho_{air}^f}{\Delta M_{cyl}}\right]^2 \left[\frac{u(\rho_{air}^f)}{\rho_{air}^f}\right]^2$$
$$+ \left(\frac{M_{sen}}{\Delta M_{cyl}}\right)^2 \left(O_{DS}^f\left[1 - \frac{\rho_{air}^f}{\rho_{ref}}\right] - O_{DS}^i\left[1 - \frac{\rho_{air}^i}{\rho_{ref}}\right]\right)^2 \left[\frac{u(M_{sen})}{M_{sen}}\right]^2 \quad (21)$$
$$+ \left(\left[\frac{M_{ref}^f + O_{DS}^f M_{sen}}{\Delta M_{cyl}}\right]\left(\frac{\rho_{air}^f}{\rho_{ref}}\right) - \left[\frac{M_{ref}^i + O_{DS}^i M_{sen}}{\Delta M_{cyl}}\right]\left(\frac{\rho_{air}^i}{\rho_{ref}}\right)\right)^2 \left[\frac{u(\rho_{ref})}{\rho_{ref}}\right]^2$$
$$+ \left(\frac{V_{cyl}(\rho_{air}^f - \rho_{air}^i)}{\Delta M_{cyl}}\right)^2 \left[\frac{u(V_{cyl})}{V_{cyl}}\right]^2 + \left[\left(\frac{M_{ref}^i}{\Delta M_{cyl}}\right)\left(1 - \frac{\rho_{air}^i}{\rho_{ref}}\right)\right]^2 \left[\frac{u(M_{ref}^i)}{M_{ref}^i}\right]^2$$
$$+ \left[\left(\frac{M_{ref}^f}{\Delta M_{cyl}}\right)\left(1 - \frac{\rho_{air}^f}{\rho_{ref}}\right)\right]^2 \left[\frac{u(M_{ref}^f)}{M_{ref}^f}\right]^2 + 2\left[\frac{M_{scale}^{res}}{\Delta M_{cyl}}\right]^2 + \sigma_{\Delta M_{cyl}}^2$$

where the last term is the standard deviation of repeated measurements (consisting of the standard deviation of the repeated initial mass measurements divided by the average initial mass root-sum-squared with the standard deviation of final mass measurements divided by the average final mass) and the second to last term accounts for the resolution of the scale. The square of the sensitivity coefficient is two in this term to account for the uncertainty of scale resolution for both the initial and final weighings. The various uncertainty components are itemized in Table 12. The total relative standard uncertainty for the mass of gas transferred into the collection tank is $[u(\Delta M_{cyl})/\Delta M_{cyl}] = 28 \times 10^{-6}$. The value of uncertainty for each component is documented below.

5.6.5.1 <u>Initial and Final Reference Masses</u> The set of NIST traceable reference masses ranges in value from 0.001 kg to 45 kg. The uncertainty differs for each mass in the set. For the combination of masses used for the initial weighing the relative standard uncertainty is $[u(M_{ref}^i)/M_{ref}^i] = 0.69 \times 10^{-6}$, and for the final weighing it is $[u(M_{ref}^f)/M_{ref}^f] = 0.75 \times 10^{-6}$. We made no effort to account for the decrease in uncertainty attributed to correlations between any common reference masses used for both the initial and final weighings. The relatively small impact of the mass measurement in the volume uncertainty warrants this simplified approach (see Table 11).

5.6.5.2 <u>Sensitivity Mass</u> The same sensitivity mass is used for both the initial and final weighings. The sensitivity mass is NIST traceable and its relative standard uncertainty is $[u(M_{sen})/M_{sen}] = 0.04 \times 10^{-6}$.

5.6.5.3 Density of the Reference and Sensitivity Masses Both reference and sensitivity masses are made of stainless steel and have identical densities equal to 7950 kg/m^3. The relative standard uncertainty for density of the reference mass (or sensitivity mass) is conservatively estimated to be $[u(\rho_{ref})/\rho_{ref}] = 9200 \times 10^{-6}$.

Table 12. Uncertainty of the mass of gas transferred into the collection tank.

Mass Transfer Uncertainty	Rel. Std. Unc. (k=1) ($\times 10^{-6}$)	Sen. Coeff. (-----)	Perc. Contrib. (%)	Unc. Type	Comments
Mass Transfer, ΔM_{cyl} = 30.0710 kg					
Initial ref. mass, M_{ref}^i = 345.7532 kg	0.69	11.5	8.1	B	See Section 5.6.5.1
Final ref. mass, M_{ref}^f = 315.6720 kg	0.75	-10.5	8.1	B	See Section 5.6.5.1
Sensitivity mass, M_{sen} = 1.0000454 kg	0.04	-2.63×10^{-4}	0.0	B	See Section 5.6.5.2
Density of ref. mass, ρ_{ref} = 7950 kg/m^3	9200	-1.63×10^{-4}	0.3	B	See Section 5.6.5.3
External vol. of cylinder, V_{cyl} = 0.27 m^3	83942	-6.72×10^{-5}	4.1	B	See Section 5.6.4.4
Initial air density, ρ_{air}^i = 1.194 kg/m^3	844	-8.79×10^{-3}	7.1	A, B	See Section 5.5.4.5
Final air density, ρ_{air}^f = 1.186 kg/m^3	616	9.26×10^{-3}	4.2	A, B	See Section 5.6.5.5
Scale Resolution, M_{scale}^{res} = 0.2 g	6.7	1.41	11.4	A	Cal. records Expts.
Std. dev. of repeated measurements	21	1	56.7	A	Three or more meas.
Combined Uncertainty	28		100		

5.6.5.4 External Cylinder Volume The measurement of external volume of the array of cylinders required only marginal accuracy since its sensitivity coefficient is small. The small sensitivity coefficient results because the buoyant forces are nearly identical during the initial and final cylinder weighings. As observed in Eqn. (20), when the air density is the same during the initial and final cylinder weighings, the buoyancy forces completely cancel. Since only marginal accuracy is necessary, this volume is measured using a tape measure. The external cylinder volume is estimated to be 0.27 m^3, and the relative standard uncertainty is $[u(V_{cyl})/V_{cyl}] = 83942 \times 10^{-6}$ or 8.4 %.

5.6.5.5 Initial and Final Air Density for Buoyancy Correction The air density was determined by measuring the pressure, temperature, and relative humidity in conjunction with the curve fit [21]

$$\rho_{air} = \left(\frac{b_1}{T}\right)(P - b_2 RH P_{sat}) \qquad (22)$$

where $P_{sat} = 1.7526 \times 10^{11} \exp(-5315.56/T)$ is the saturation pressure, and the values of the coefficients are $b_1 = 0.0034848$ K·Pa^{-1}kg/m^3 and $b_2 = 0.003796$ respectively. For relative humidities between 0 percent and 100 percent, at ambient pressures, and temperatures ranging from 290 K to 300 K, the relative difference between the densities calculated using Eqn. (22) and the Refprop thermodynamic database was less than 300×10^{-6}. Assuming a rectangular distribution, the relative standard uncertainty of Eqn. (22) is 173×10^{-6}. The additional uncertainty components for the initial air density are shown in Table 13, and the total relative standard uncertainty is $[u(\rho_{air}^i)/\rho_{air}^i] = 844 \times 10^{-6}$. The major sources of uncertainty are attributed to the measurement of temperature and relative humidity. The temperature uncertainty is primarily due to temperature drift in the room during the weighing procedure, and the relative humidity uncertainty results from uncertainty in its calibration as well as instrument drift during the measurement. The uncertainty components for the final air density have values comparable to the corresponding initial values, and corresponding components have the same uncertainty type. The total relative standard uncertainty for the final air density is $[u(\rho_{air}^f)/\rho_{air}^f] = 616 \times 10^{-6}$.

Table 13. Uncertainty of the initial air density for buoyancy correction.

Initial Air Density Uncertainty	Rel. Std. Unc. (k=1)	Sen. Coeff.	Perc. Contrib.	Unc. Type	Comments
Initial air density, $\rho_{air}^i = 1.194$ kg/m³	($\times 10^{-6}$)	(-----)	(%)	(A, B)	
Equation of state for moist air	173	1	4.2	B	Comparison w/ Refprop [10]
Relative humidity, $RH_{air}^i = 20.9$ percent	169750	-1.8×10^{-3}	14.0	B	Cal. residuals of RH sensor
Initial room temperature, $T_{air}^i = 293.24$ K	712	-1.0	76.2	A, B	Mainly temp. drift during meas.
Initial room pressure, $P_{air}^i = 100.654$ kPa	200	1.0	5.6	A, B	Mainly pres. drift during meas.
Combined Uncertainty	**844**		**100**		

6. *PVTt* Mass Flow Uncertainty

At the most basic level, a *PVTt* flow measurement system involves measuring mass and time. In particular, the accumulated mass in the inventory volume and in the collection tank is measured over the collection period. For a general *PVTt* system, the time-averaged CFV mass flow was given previously in Eqn.(8), but repeated here for convenience

$$\dot{m} = \frac{\Delta M_T + \Delta M_I}{\Delta t} \tag{8}$$

where the uncertainty attributed to leaks is omitted in the calculation of mass flow but considered in the uncertainty analysis. The expression for the mass flow uncertainty is determined by applying the method of propagation of uncertainty to Eqn. (8). The resulting expression of uncertainty is

$$\left[\frac{u(\dot{m})}{\dot{m}}\right]^2 = \left(\frac{\Delta M_I}{\Delta M}\right)^2 \left[\frac{u(\Delta M_I)}{\Delta M_I}\right]^2 + \left(\frac{\Delta M_T}{\Delta M}\right)^2 \left[\frac{u(\Delta M_T)}{\Delta M_T}\right]^2 + \left[\frac{u(\Delta t)}{\Delta t}\right]^2 + \left[\frac{u(\dot{m}_{leak})}{\dot{m}_{leak}}\right]^2 \\ + 2\left(\frac{\Delta M_I}{\Delta M}\right)\left(\frac{\Delta M_T}{\Delta M}\right)\left(\left[\frac{u(\mathcal{M})}{\mathcal{M}}\right]^2 + \left[\frac{u(R_u)}{R_u}\right]^2\right) + \sigma_{ss}^2 \tag{23}$$

where the last term is the uncertainty attributed to the steady state assumption, and the second to last term accounts for the correlation between $u(\Delta M_I)$ and $u(\Delta M_T)$ attributed to the parameters $\mathcal{M}$ and R_u which are common to both ΔM_I and ΔM_T (*i.e.*, see Eqns. 7a and 7b). Since the inventory mass cancellation technique (section 4.5) ensures that ΔM_I is identically zero, the relative uncertainty of $[u(\Delta M_I)/\Delta M_I]$ is infinite, making it impractical to use. By canceling the repeated occurrences of ΔM_I in the numerator and denominator of the first term and observing that the correlated terms vanish as ΔM_I tends to zero we obtain

$$\left[\frac{u(\dot{m})}{\dot{m}}\right]^2 = \left[\frac{u(\Delta M_I)}{\Delta M_T}\right]^2 + \left[\frac{u(\Delta M_T)}{\Delta M_T}\right]^2 + \left[\frac{u(\Delta t)}{\Delta t}\right]^2 + \left[\frac{u(\dot{m}_{leak})}{\dot{m}_{leak}}\right]^2 + \sigma_{ss}^2 \quad (24)$$

the appropriate mass flow uncertainty corresponding to the inventory matching technique where we note that inventory mass uncertainty is normalized by ΔM_T and the sensitivity coefficient for the tank mass uncertainty is unity (i.e., $\Delta M_T = \Delta M$).

Table 14. Uncertainty of the CFV mass flow.

PVTt Mass Flow Uncertainty	Rel. Std. Unc. (k=1)	Sen. Coeff.	Perc. Contrib.	Unc. Type	Comments
CFV mass flow, $\dot{m}$ =30.0710 kg	($\times 10^{-6}$)	(-----)	(%)	(A, B)	
Tank mass accumulation, ΔM_T =29.2054 kg	379	1	73.1	A, B	See Section 6.1
Inv. Vol. mass accumulation, ΔM_I = 0 kg	206	1	21.7	A, B	See Section 6.2
Collection time, Δt = 20 s	15	1	0.1	A, B	See Section 5.2
Steady state assumption	100	1	5.1	B	See Section 6.4
Leaks	0	1	0.0	B	See Section 6.3
Combined Uncertainty	**443**		**100**		

The five uncertainty components shown in Eqn. (24) are itemized in Table 14. Although the inventory mass cancellation technique ensures that $\Delta M_I = 0$, as observed in the table, its uncertainty is not zero. The most significant contribution to the mass flow uncertainty (i.e., more than 73 %) stems from measuring the mass accumulated in the collection tank. When Eqn. (24) is used to combine all of the uncertainty components shown in the table, the relative standard uncertainty for mass flow is $[u(\dot{m})/\dot{m}] = 443 \times 10^{-6}$, and the expanded relative uncertainty is $[U_{exp}(\dot{m})/\dot{m}] = 0.09\%$ (i.e., $k = 2$). A detailed explanation of each uncertainty component is given below.

6.1 Accumulated Mass in the Collection Tank

The mass accumulation in the collection tank is determined volumetrically using Eqn. (7a) derived in section 4.2 and repeated here for convenience

$$\Delta M_T = (\mathcal{M}/R_u)\left(\frac{P_T^f}{Z_T^f T_T^f} - \frac{P_T^i}{Z_T^i T_T^i}\right)V_T. \quad (7a)$$

The corresponding uncertainty is

$$\left[\frac{u(\Delta M_T)}{\Delta M_T}\right]^2 = \left[\frac{u(\mathcal{M})}{\mathcal{M}}\right]^2 + \left[\frac{u(R_u)}{R_u}\right]^2 + \left(\frac{M_T^i}{\Delta M_T}\right)^2\left[\left(\frac{u(P_T^i)}{P_T^i}\right)^2 + \left(\frac{u(T_T^i)}{T_T^i}\right)^2 + \left(\frac{u(Z_T^i)}{Z_T^i}\right)^2\right] \quad (25)$$

$$+\left(\frac{M_T^f}{\Delta M_T}\right)^2\left[\left(\frac{u(P_T^f)}{P_T^f}\right)^2+\left(\frac{u(T_T^f)}{T_T^f}\right)^2+\left(\frac{u(Z_T^f)}{Z_T^f}\right)^2\right]+\left[\frac{u(V_T)}{V_T}\right]^2$$

and each component is itemized in Table 15. The uncertainties in the initial pressure, temperature, and compressibility factor play a reduced role in the uncertainty analysis since their corresponding sensitivity coefficients are much less than unity. More than 86 % of the uncertainty results from determining the tank volume and the average final tank temperature. Details for the uncertainty of each of these components are located in sections 5.1 to 5.5. The total relative standard uncertainty for the mass accumulated in the collection tank is $[u(\Delta M_T)/\Delta M_T] = 379 \times 10^{-6}$.

Table 15. Uncertainty of the mass accumulation in the collection tank.

Uncertainty of mass accumulation in the tank	Rel. Std. Unc. (k=1) (×10⁻⁶)	Sen. Coeff. (-----)	Perc. Contrib. (%)	Unc. Type (A, B)	Comments
Tank mass accumulation, ΔM_T = 29.2054 kg					
Tank volume, V_T =25.8969 m³	276	1	53.0	A, B	See Table 11
Tank initial temp., T_T^i =293 K	4115	0.001	0.0	A, B	See Section 5.5
Tank final temp., T_T^f =294 K	220	-1.001	33.7	A, B	See Table 10
Tank Initial pres., P_T^i =0.1 kPa	6120	0.001	0.0	B	See Table 8
Tank final pres., P_T^f =95.246 kPa	118	-1.001	9.7	A, B	See Table 9
Tank initial compressibility., Z_T^i =1	50	0.001	0.0	B	See Section 5.1.3
Tank final compressibility., Z_T^f =1	50	-1.001	1.8	B	See Section 5.1.3
Molecular mass, $\mathcal{M}_{air}$ =28.9647 g/mol	51	1	1.8	A, B	See Section 5.1.2
Univ. gas const., R_{univ} =8314.472 J/kmol·K	1.7	-1	0.0	B	See Section 5.1.1
Combined Uncertainty	**379**		**100**		

6.2 Accumulated Mass in the Inventory Volume

As previously defined in Eqn. (6b), and repeated here for convenience, the mass accumulated in the inventory volume equals the density change in this volume multiplied by the size of the inventory volume

$$\Delta M_I = \Delta \rho_I V_I. \tag{6b}$$

where $\Delta \rho_I \equiv \rho_I^f - \rho_I^i$ is the density change between the initial and final density. Using the method of propagation of uncertainty, the uncertainty in the mass accumulated in the inventory volume is

$$\left[\frac{u(\Delta M_I)}{\Delta M_T}\right]^2 = \left(\frac{V_I}{V_T}\right)^2\left[\frac{u(\Delta \rho_I)}{\Delta \rho_T}\right]^2 + \left(\frac{\Delta M_I}{\Delta M_T}\right)^2\left[\frac{u(V_I)}{V_I}\right]^2 \tag{26}$$

where the inventory mass cancellation technique ensures that the second term is identically zero (i.e., $\Delta M_I = 0$). The uncertainty attributed to the density change in the inventory volume is normalized by the density change in the collection tank, $\Delta\rho_T$, instead of $\Delta\rho_I=0$, to avoid the singularity that would result from dividing by zero. Table 16 itemizes these components and shows that the relative standard uncertainty for the inventory volume mass accumulation is $[u(\Delta M_I)/\Delta M_T] = 206 \times 10^{-6}$. Since the size of the inventory volume contributes no uncertainty, all of the uncertainty derives from the density change.

Table 16. Uncertainty components for the inventory volume mass accumulation.

Uncertainty of Inv. Vol. Density Difference	Rel. Std. Unc. (k=1)	Sen. Coeff.	Perc. Contrib.	Unc. Type	Comments
Mass accumulation in inv. vol., ΔM_I =0 kg	(%)	(-----)	(%)	(A, B)	
Density diff. in Inv. Vol., $\Delta\rho_I$ =0 kg/m^3	68162	0.003	100	A, B	See Section 6.2
Inv vol. size, V_I =0.078 m^3	250000	0	0	B	See Section 5.4
Combined Uncertainty			206	100	

The uncertainty of the density difference in the inventory volume, $\Delta\rho_I$, is less than the uncertainty of either ρ_I^i or ρ_I^f individually. The lower uncertainty results from the cancellation of correlated sources of uncertainty between the initial and final inventory volume densities. The correlated sources between ρ_I^i or ρ_I^f can be evaluated in a straight forward manner if the density difference is expressed in terms of pressures, temperatures, etc. as given previously in Eqn.(7b)

$$\Delta\rho_I = (\mathcal{M}/R_u)\left(\frac{P_I^f}{Z_I^f T_I^f} - \frac{P_I^i}{Z_I^i T_I^i}\right). \tag{7b}$$

By applying the propagation of uncertainty the uncertainty of the density difference is

$$\left[\frac{u(\Delta\rho_I)}{\Delta\rho_T}\right]^2 = [u_{tot}(P)]_{Inv}^2 + [u_{tot}(T)]_{Inv}^2 + [u_{tot}(Z)]_{Inv}^2 + \left(\frac{\Delta\rho_I}{\Delta\rho_T}\right)^2\left[\left[\frac{u(\mathcal{M})}{\mathcal{M}}\right]^2 + \left[\frac{u(R_u)}{R_u}\right]^2\right] \tag{27}$$

where the inventory mass cancellation technique ensures that the last term is zero. The first three terms are the total uncertainty for the initial and final pressure measurements, the total uncertainty of the temperature measurements, and the total uncertainty of the compressibility factor, respectively. The uncertainty of each of these terms is discussed here.

6.2.1 Total Pressure Uncertainties in the Inventory Volume

The total pressure uncertainty in the inventory volume is

$$[u_{tot}(P)]_{Inv}^2 = \left(\frac{\rho_I^i}{\Delta\rho_T}\right)^2\left[\left(\frac{u_c(P_I^i)}{P_I^i} - \frac{u_c(P_I^f)}{P_I^f}\right)^2 + \left(\frac{u_u(P_I^i)}{P_I^i}\right)^2 + \left(\frac{u_u(P_I^f)}{P_I^f}\right)^2\right] \tag{28}$$

where the initial correlated and uncorrelated uncertainties are given in section 5.3.1 as $[u_c(P_I^i)/P_I^i] = 4.0\%$ and $[u_u(P_I^i)/P_I^i] = 2.3\%$ respectively, and the final correlated and uncorrelated uncertainties are given in section 5.3.2. as $[u_c(P_I^f)/P_I^f] = 4.1\%$ and $[u_u(P_I^f)/P_I^f] = 2.3\%$ respectively. The total pressure uncertainty is $[u_{tot}(P)]_{Inv} = 5.5\%$.

Table 17. Uncertainty components for the inventory volume density difference.

Uncertainty of Density Difference	Rel. Std. Unc. ($k=1$) (%)	Sen. Coeff. (-----)	Perc. Contrib. (%)	Unc. Type (A, B)	Comments
Density Difference, $\Delta\rho_I = 0$ kg/m^3					
Total Inv. Vol. pressure uncertainty	4.1	1	35.5	A, B	See section 6.2.1
Total Inv. Vol. temperature uncertainty	5.5	1	64.5	A, B	See section 6.2.2
Total Inv. Vol. compressibility uncertainty	0	1	0.0	B	See section 6.2.3
Molecular mass, $\mathcal{M}_{air} = 28.9647$ g/mol	0.0051	0	0.0	A, B	See section 5.1.2
Univ. gas const., $R_{univ} = 8314.472$ J/(kg·K)	0.0002	0	0.0	B	See section 5.1.1
Combined Uncertainty	**6.8**		**100**		

6.2.2. Total Temperature Uncertainties in the Inventory Volume

The total temperature uncertainty in the inventory volume is

$$[u_{tot}(T)]^2_{Inv} = \left(\frac{\rho_I^i}{\Delta\rho_T}\right)^2 \left[\left(\frac{u_c(T_I^i)}{T_I^i} - \frac{u_c(T_I^f)}{T_I^f}\right)^2 + \left(\frac{u_u(T_I^i)}{T_I^i}\right)^2 + \left(\frac{u_u(T_I^f)}{T_I^f}\right)^2\right] \quad (29)$$

where the initial correlated and uncorrelated temperature uncertainties are given in section 5.3.3. as $[u_c(T_I^i)/T_I^i] = 9.6\%$ and $[u_u(T_I^i)/T_I^i] = 1.7\%$ respectively, and the final correlated and uncorrelated uncertainties are given in section 5.3.4. as $[u_c(T_I^f)/T_I^f] = 9.7\%$ and $[u_u(T_I^f)/T_I^f] = 1.7\%$ respectively. The total relative temperature uncertainty is $[u_{tot}(T)]_{Inv} = 4.1\%$.

6.2.3 Total Compressibility Factor Uncertainties in the Inventory Volume

The uncertainty in the compressibility factor consists only of correlated uncertainty components. Any uncorrelated uncertainty components are fossilized as correlated components by the curve fit used to represent the experimental data [16]. Since the initial and final inventory thermodynamic conditions are nearly identical, these correlated components completely cancel so that the net uncertainty is zero.

6.3 Effect of Leaks

The influences of leaks on *PVTt* flow measurements are most pronounced at the lowest flows (200 L/min). At low flows, the collection time is longer so that the sub-atmospheric pressures in the collection tank and inventory volume persist for a longer duration, and leaks makeup a larger fraction of the accumulated mass. To avoid this situation, the FMG inspects its flow standard for leaks prior to each calibration. If the source of a leak cannot be identified, the size of the leak is estimated by multiplying the rate of density increase from an initially evacuated collection tank and/or inventory volume to the appropriate volume. The measured leak rate is then included in reported uncertainty of a given calibration. For the purposes of this document the uncertainty attributed to leaks is assumed to be zero.

6.4 Uncertainty Attributed to the Steady Flow Assumption

In developing the expression for the measured mass flow (*i.e.*, Eqn. 8 or 9) we assumed that the flow entering the CFV remained steady for the entire collection period. However, in practice steady flow conditions at the CFV inlet are never perfectly attained. Instead, the PID controller used to set the flow maintains pseudo steady state conditions, whereby the flow fluctuates about a fixed baseline. Here we propose a conservative method for estimating the uncertainty associated with these fluctuations.

The mass flow through a choked CFV under steady flow conditions is [22]

$$\dot{m}_{CFV} = \frac{P_o A_t C_s C_d \sqrt{M}}{\sqrt{R_u T_o}} \tag{30}$$

where P_o is the stagnation pressure, T_o is the stagnation temperature, A_t is the throat area, C_s is the critical flow function, and C_d is the discharge coefficient. Steady flow conditions are obtained by maintaining both P_o and T_o constant throughout the collection period. However, Eqn. (30) can still be used under pseudo steady state conditions (*i.e.*, small fluctuations in P_o and T_o) if $\dot{m}_{CFV}$ is time-averaged over the collection period. In this case, an estimate of the uncertainty attributed to unsteady effects is taken to be the standard deviation of the mass flow over the collection period ($\sigma_{\dot{m}_{CFV}}$). A typical value for the relative standard uncertainty attributed to unsteady effects is [$\sigma_{\dot{m}_{CFV}} / \dot{m}_{CFV}$] = 100×10^{-6}.

In calculating both the average mass flow and its standard deviation we assume that C_d is unaffected by small fluctuations in P_o and T_o. Physically, the discharge coefficient corrects for boundary layer effects along the CFV wall and for curvature of the sonic line near the CFV throat [23]. A small change in either P_o or T_o does not significantly alter the thickness of the boundary layer or change the shape of the sonic line so that the changes in C_d are of second order. These second order effects can be neglected when assessing the uncertainty attributed to steady state assumption.

7. Summary

This document addresses the flow measurement capabilities of the 26 m³ *PVTt* system, the United States primary standard for measuring the flow of dry air. Flow measurements are conducted at ambient temperature and at pressures up to 800 kPa, and the flow range extends from 200 standard L/min to 77000 standard L/min. This document explains the function of the various components comprising the *PVTt* system, develops the theoretical basis for *PVTt* mass flow measurements, explains the underlying principles for its operation, spells out the operating procedures used for flowmeter calibrations, provides details necessary for customers wanting to submit a meter for calibration (*i.e.*, pipeline sizes and available fittings, cost, turnaround time, etc.), and gives a detailed analysis of the uncertainty of mass flow measurements.

The uncertainty for mass flow is assessed using the method of propagation of uncertainty [16]. The analysis shows that the expanded uncertainty of mass flow is 0.09 % with a coverage factor of two. The various uncertainty components are itemized in sections 5 and 6. For convenience a summary of the primary uncertainty components is given in Table 18. The largest components of uncertainty are attributed measuring the collection tank volume and the final (*i.e.*, after filling) temperature of the gas in the tank in the collection tank. Together these contribute more than 60 % of the overall uncertainty. Any future uncertainty reductions are likely to focus on improving the accuracy of these measurements.

Table 18. Uncertainty of the *PVTt* mass flow.

PVTt Mass Flow Uncertainty	Rel. Std. Unc. (*k*=1)	Sen. Coeff.	Perc. Contrib.	Unc. Type	Comments
CFV mass flow, $\dot{m}$ =30.0710 kg	($\times 10^{-6}$)	(-----)	(%)	(A, B)	
Collection Tank Uncertainties					
Tank volume, V_T = 25.8969 m^3	276	1	38.7	A, B	See Table 11
Tank initial temp., T_T^i = 293 K	4115	0.001	0.0	A, B	See Section 5.5
Tank final temp., T_T^f = 294 K	220	-1.001	24.7	A, B	See Table 10
Tank initial pres., P_T^i = 0.1 kPa	6120	-0.001	0.0	A, B	See Table 8
Tank final pres., P_T^f = 95.246 kPa	118	1.001	7.1	A, B	See Table 9
Tank initial compressibility, Z_T^i = 1	50	0.001	0.0	B	See Section 5.1.3
Tank final compressibility, Z_T^f = 1	50	-1.001	1.3	B	See Section 5.1.3
Inventory Volume Uncertainties					
Total Inv. Vol. Pres. Unc., P_I^i & P_I^f	166	1	14.0	A, B	See section 6.2.1
Total Inv. Vol. Temp. Unc., T_I^i & T_I^f	123	1	7.7	A, B	See section 6.2.2
Total Inv. Vol. Comp. Unc., Z_I^i & Z_I^f	0	1	0.0	B	See section 6.2.3
Inv vol. size., V_I =0.078 m^3	250000	0	0.0	B	See Section 5.4
Reference Properties					
Molecular mass, $\mathcal{M}_{air}$ =28.9647 g/mol	51	1	1.3	A, B	See Section 5.1.2
Univ. gas const., R_{univ} =8314.472 J/kmol·K	1.7	-1	0.0	B	See Section 5.1.1
Timing Uncertainties					
Collection time, Δt = 20 s	15	1	0.1	A, B	See Section 5.2
Unsteady Effect and Leaks					
Steady state assumption	100	1	5.1	B	See Section 6.4
Leaks	0	1	0.0	B	See Section 6.3
Combined Uncertainty	**443**		**100**		

The 26 m^3 *PVTt* system and both of its smaller counterparts (*i.e.*, the 34 L and the 677 L *PVTt* flow standards) all implement an inventory mass cancellation technique to reduce the uncertainty of the dynamic pressure and temperature measurements made in the inventory volume. The technique works by ensuring pre-filling and after-filling thermodynamic conditions in the inventory volume are nearly identical. In this way many of the correlated uncertainties associated with using the same instrumentation to measure nearly the same conditions cancel each other, thereby making little or no contribution to the overall uncertainty. Finally, all three *PVTt* systems are completely automated and able to perform calibrations overnight and on weekends, thereby expediting turnaround time for our customers.

REFERENCES

[1] International Organization for Standardization, *International Vocabulary of Basic and General Terms in Metrology*, 2nd edition, 1993.

[2] Brain T. J. S., Macdonald, L. M., *Evaluation of the Performance of Small Scale Critical Flow Venturis Using the NEL Gravimetric Gas Flow Standard*, Technical Paper: B-3.

[3] Wright, J. D., and Mattingly, G. E., *NIST Calibration Services For Gas Flow Meters: Piston Prover and Bell Prover Gas Flow Facilities*, NIST SP 250-49, 1998.

[4] Todd, D. A., *NPSL Method for Calibrating Bell Provers*, Technical Report: Navy Primary Standards Laboratory Code 4.1.4.5.0, 1996.

[5] Olsen, L. and Baumgarten, G., *Gas Flow Measurement by Collection Time and Density in a Constant Volume*, Flow: Its Measurement and Control in Science and Industry, ISA, (1971), pp. 1287 - 1295.

[6] Wright, J. D., Johnson, A. N., and Moldover, M. R., *Design and Uncertainty Analysis for a PVTt Gas Flow Standard*, J. Res. Natl. Inst. Stand. Technology 108, 21-47 (2003)

[7] Wright, J. D., *What Is the "Best" Transfer Standard for Gas Flow?"*, FLOMEKO, Groningen, the Netherlands, May, 2003.

[8] Wright, J. D., *The Long Term Calibration Stability of Critical Flow Nozzles and Laminar Flowmeters*, National Conference of Standards Laboratories Conference Proceedings, Albuquerque, NM, USA, pp. 443-462, 1998.

[9] Marshall, J. L., *NIST Calibration Services Users Guide 1998*, NIST Special Publication 250, January, 1998.

[10] Lemmon, E. W., McLinden, M. O., and Huber, M. L., *Refprop 23: Reference Fluid Thermodynamic and Transport Properties, NIST Standard Reference Database 23, Version 7*, National Institute of Standards and Technology, Boulder, Colorado, 2002.

[11] Moldover, M. R., Trusler, J. P. M., Edwards, T. J., Mehl, J. B., and Davis, R. S., *Measurement of the Universal Gas Constant R Using a Spherical Acoustic Resonator*, NIST J. of Res., 93, (2), 85–143, 1988.

[12] B. N. Taylor and C. E. Kuyatt, *Guidelines for the Evaluating and Expressing the Uncertainty of NIST Measurements Results*, NIST TN-1297, Gaithersburg, MD: NIST 1994.

[13] J. Hilsenrath, C. W. Beckett, W. S. Benedict, L. Fano, H. J. Hoge, J. F. Masi, R. L. Nuttall, Y. S. Touloukian, H. W. Wooley, *Tables of Thermal Properties of Gases*, U.S. Department of Commerce NBS Circular 564, 1955.

[14] R. C. Weast, *CRC Handbook of Chemistry and Physics,* CRC Press Inc., 58th edition, Ohio, 1977.

[15] F. T. Mackenzie and J. A. Mackenzie, *Our changing planet.* Prentice-Hall, Upper Saddle River, NJ, p 288-307, 1995 (After Warneck, 1988; Anderson, 1989; Wayne, 1991.)

[16] Coleman, H. W. and W. G. Steele, *Experimentation and Uncertainty Analysis for Engineers*, John Wiley and Sons, 2nd edition, 1999.

[17] Wright J. D. and Johnson, A. N., *Uncertainty in Primary Gas Flow Standards Due to Flow Work Phenomena*, FLOMEKO, Salvador, Brazil (2000).

[18] A. Bejan, *Convection Heat Transfer*, John Wiley and Sons, 1st edition 1984.

[19] A. N. Johnson, J. D. Wright, M. R. Moldover, P. I. Espina, *Temperature Characterization in the Collection Tank of the NIST 26 m^3 PVTt Gas Flow Standard*, Metrologia, 2003, 40, 211-216.

[20] E. W. Lemmon, R. T. Jacobsen, S. G. Penoncello, and D. G. Friend, *Thermodynamic Properties of Air and Mixtures of Nitrogen, Argon, and Oxygen from 60 to 2000 K at Pressures to 2000 MPa*, J. Phys. Chem. Ref. Data 29, (3), 331-362, (2000).

[21] Wagner, W. and Pruss, A., *The IAPWS Formulation 1995 for the Thermodynamic Properties of Ordinary Water Substance for General and Scientific Use*, J. Phys. Chem. Ref. Data, 31(2): 387-535, 2002.

[22] ISO 9300: (E)., *Measurement of Gas Flow by Means of Critical Flow Venturi Nozzles*, Geneva Switz, 1990.

[23] Johnson, A. N., *Numerical Characterization of the Discharge Coefficient in Critical Nozzles*, Ph.D. Dissertation, Pennsylvania State Univ., College Park, PA, 2000.

SAMPLE CALIBRATION REPORT

FOR

A CRITICAL FLOW NOZZLE

July 22, 2005

Mfg.: CFV Builders, Inc.
Serial Number: 1234
Throat Diameter: 0.19009 inch (0.48283 cm)

submitted by

Flow Nozzles, Inc.
Metertown, MD

Purchase Order No. A123 dated May 24, 2004

The flow meter identified above was calibrated by flowing filtered dry air at a constant rate through it into a volumetric prover (the NIST 26 m^3 *PVTt* standard). The *PVTt* standard determines mass flow, $\dot{m}$, by measuring the change in density of gas diverted into a known volume for a measured period of time.[1] The flow meter was tested at five flows and at each flow, three (or more) measurements were gathered on two different occasions and used to produce averages at each of these flows. As a result, the tabulated data for this test are averages of six or more individual calibration measurements.

A photograph of the flow meter installation is shown in Figure 1. The nozzle temperature, (T_1), and pressure, (P_1), were measured with NIST sensors, (Keithley SN 687848, thermistor #26, and Paroscientific SN 73965). Stagnation temperature, T_0, was calculated from the measured temperature via the following equation, using a recovery factor, r, of 0.75:

$$T_0 = T_1 \cdot \left[1 + \frac{\gamma - 1}{2} \cdot M^2 \cdot r \right] \qquad (1)$$

and the stagnation pressure, P_0, was calculated via the equation:

---

[1] Johnson, A.N., Wright, J.D., Moldover, M.R., and Espina, P.I., *Temperature Characterization in the collection tank of the NIST 26 m^3 PVTt Gas Flow Standard*, Metrologia, 40, 211-216, 2003.

$$P_0 = P_1 \cdot \left[1 + \frac{\gamma-1}{2} \cdot M^2\right]^{\frac{\gamma}{\gamma-1}} \tag{2}$$

where γ is the specific heat ratio and M is the Mach number in the approach pipe ($d = 3.48$ cm), both based on P_1 and T_1.[2] The largest of these corrections is 0.05 % for pressure.

Figure 1. Photograph of the flow meter installation.

The Reynolds number is included in the tabulated data and it was calculated using the following expression:

$$Re = \frac{4 \cdot \dot{m}}{\pi \cdot d \cdot \mu} \tag{3}$$

where $\dot{m}$ is the mass flow of gas, d is the nominal nozzle throat diameter, and μ is the gas viscosity, all in consistent units so that Re is dimensionless. The gas properties (density and viscosity) were calculated using best-fit equations which are based on the NIST gas properties

---

[2] *Measurement of Gas Flow by Means of Critical Flow Venturi Nozzles*, ISO 9300: 1990 (E), International Organization for Standardization, Geneva, Switzerland, 1990.

SAMPLE CALIBRATION REPORT
Flow Nozzles, Inc.

database.[3,4] In January 2003, the correlation for viscosity used by the NIST Fluid Flow Group was changed from an older reference to the one used in this report.

The discharge coefficient C_d was calculated from the expression:

$$C_d = \frac{4 \cdot \dot{m} \cdot \sqrt{R \cdot T_0}}{\pi \cdot d^2 \cdot P_0 \cdot C^*} \qquad (4)$$

where R is the gas constant [the universal gas constant, 8.314471 J / (mol K), divided by the gas molecular weight, 28.9646 g/mol]. The critical flow factor, C^*, was calculated from the expression:

$$C^* = \sqrt{\gamma \left(\frac{2}{\gamma+1}\right)^{\frac{\gamma+1}{\gamma-1}}} \qquad (5)$$

where γ is the specific heat ratio.

The calibration results are presented in the following table and figure. The figure shows the discharge coefficient as a function of the inverse square-root of the Reynolds number. For many ISO standardized nozzles in the laminar flow range[5] this has the effect of linearizing the calibration data.

---

[3] Lemmon, E.W., McLinden, M.O., and Huber, M.L., Refprop 23: Reference Fluid Thermodynamic and Transport Properties, NIST Standard Reference Database 23, Version 7, National Institute of Standards and Technology, Boulder, Colorado, 2002.

[4] Wright, J., Gas Properties Equations for the NIST Fluid Flow Group Gas Flow Measurement Calibration Services, 2/04.

[5] *Measurement of Gas Flow by Means of Critical Flow Venturi Nozzles*, ISO 9300: 1990 (E), International Organization for Standardization, Geneva, Switzerland, 1990.

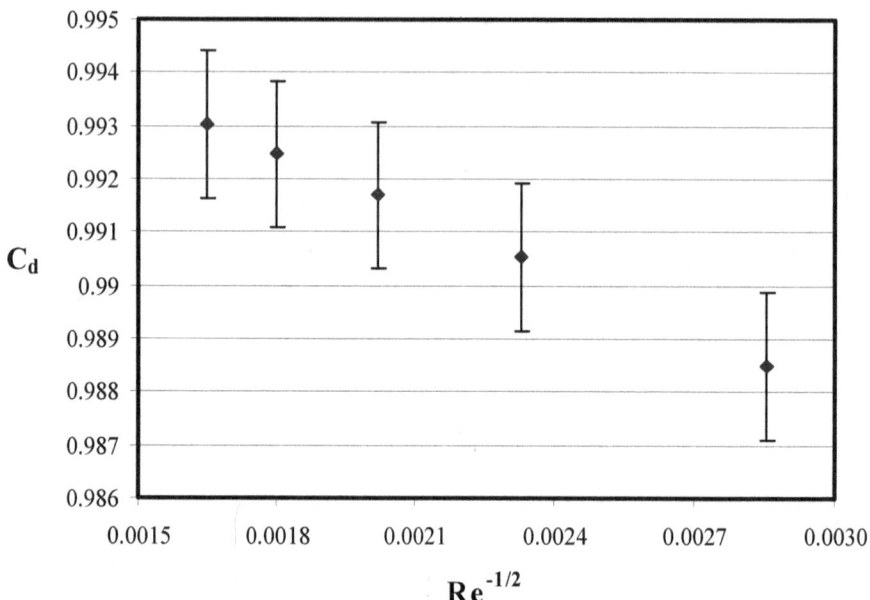

Figure 2. Calibration results for 0.19009 in (0.48283 cm), NIST xxx nozzle.

Table 1. Calibration results for 0.19009 in (0.48283 cm), NIST nozzle.

P_0 [kPa]	T_0 [K]	$\dot{m}$ [g/s]	C^* [ ]	Re [ ]	C_d [ ]	U [%]
200.44	294.72	8.5482	0.6854	1.229×10^5	0.9885	0.11
300.48	295.32	12.8340	0.6857	1.841×10^5	0.9905	0.11
400.53	295.60	17.1264	0.6860	2.452×10^5	0.9917	0.11
500.58	294.77	21.4613	0.6863	3.077×10^5	0.9925	0.11
601.63	296.26	25.7546	0.6866	3.684×10^5	0.9930	0.11

An analysis was performed to assess the uncertainty of the results obtained for the meter under test.[6,7,8] The process involves identifying the equations used in calculating the calibration result

---

[6] International Organization for Standardization, *Guide to the Expression of Uncertainty in Measurement*, Switzerland, 1996 edition.

(measurand) so that the sensitivity of the result to uncertainties in the input quantities can be evaluated. The approximately 67 % confidence level uncertainty of each of the input quantities is determined, weighted by its sensitivity, and combined with the other uncertainty components by root-sum-square to arrive at a combined uncertainty (u_c). The combined uncertainty is multiplied by a coverage factor of 2.0 to arrive at an expanded uncertainty (U) of the measurand with approximately 95% confidence level.

As described in the references, if one considers a generic basis equation for the measurement process, which has an output, y, based on N input quantities, x_i,

$$y = y(x_1, x_2, \ldots, x_N) \tag{6}$$

and all uncertainty components are uncorrelated, the normalized expanded uncertainty is given by,

$$\frac{U_e(y)}{y} = k\frac{U_c(y)}{y} = k\sqrt{\sum_{i=1}^{N} s_i^2 \left(\frac{u(x_i)}{x_i}\right)^2} \tag{7}$$

In the normalized expanded uncertainty equation, the $u(x_i)$'s are the standard uncertainties of each input, and s_i's are their associated sensitivity coefficients, given by,

$$s_i = \frac{\partial y}{\partial x_i} \frac{x_i}{y} \tag{8}$$

The normalized expanded uncertainty equation is convenient since it permits the usage of relative uncertainties (in fractional or percentage forms) and of dimensionless sensitivity coefficients. The dimensionless sensitivity coefficients can often be obtained by inspection since for a linear function they have a magnitude of unity.

For this calibration, the uncertainty of the discharge coefficient has components due to the measurement of the mass flow by the primary standard, $u(\dot{m}) = 0.06\%$,[9] as well as the pressure, $u(P) = 0.02\%$, and temperature, $u(T) = 0.03\%$, measurements at the meters under test. The sensitivity coefficients for mass flow and pressure are 1, and the sensitivity coefficient for temperature is ½. This uncertainty analysis assumes that the user will use the same values for the throat diameter and the critical flow factor given herein and that the measurement errors in these quantities are correlated and cancel.

---

[7] Taylor, B.N. and Kuyatt, C.E., *Guidelines for Evaluating and Expressing the Uncertainty of NIST Measurement Results*, NIST TN 1297, 1994 edition.

[8] Coleman, H.W. and Steele, W.G., *Experimentation and Uncertainty Analysis for Engineers*, John Wiley and Sons, 2nd ed., 1999.

[9] Johnson, A.N., Wright, J.D., Moldover, M.R., and Espina, P.I., *Temperature Characterization in the collection tank of the NIST 26 m^3 PVTt Gas Flow Standard*, Metrologia, 40, 211-216, 2003.

SAMPLE CALIBRATION REPORT Gas Flow Meter, S/N 1234
Flow Nozzles, Inc. Purchase Order No. A123

The present uncertainty analysis does not include uncertainty in the experimental measurements of viscosity found in the references, which can amount to 1% or more. To prevent errors due to viscosity, the user must use the same gas and viscosity expression used by NIST when using the results given in Table 1, or must use calibration coefficients calculated with their preferred viscosity relationship. Flow measurements made with this nozzle and a gas other than air (including humid air) will have greater uncertainty than that given in the present analysis due to uncertainty in the gas viscosity. Given these assumptions, the viscosity uncertainty depends primarily on the uncertainty of the gas temperature measurement.

To measure the reproducibility[10] of the test, the standard deviation of the discharge coefficient at each of the nominal flows was used to calculate the relative standard uncertainty (the standard deviation divided by the mean and expressed as a percentage). The reproducibility was propagated along with the other uncertainty components to calculate the combined uncertainty. Using the values given above, results in the expanded uncertainties listed in the data table and shown as error bars in the figure.

For the Director,
National Institute of Standards and Technology

Project Leader's Official Signature

Aaron N. Johnson
Fluid Flow Group
Process Measurements Division
Chemical Science and Technology Laboratories

---

[10] Reproducibility is herein defined as the closeness of agreement between measurements with the flow changed and then returned to the same nominal value.

www.ingramcontent.com/pod-product-compliance
Lightning Source LLC
Chambersburg PA
CBHW081905170526
45167CB00007B/3158